W9-BXX-870

Out of the Blue

Why is the sky ¹
rainbow? Wh·
the Blue, s¹
and whe·
and ho
night sky ﹍
the naked eye.
and written in a ρ
how, *Out of the Blue*
who feels curious or puz﹍

JOHN NAYLOR has been fascin﹍
child growing up in Peru, and by the
to London University to study engineer﹍
teaches physics at a secondary school in Lo﹍

Out of the Blue

A 24-hour Skywatcher's Guide

JOHN NAYLOR

CAMBRIDGE
UNIVERSITY PRESS

WITHDRAWN

URBANDALE PUBLIC LIBRARY
3520 86TH STREET
URBANDALE, IA 50322-4056

PUBLISHED BY THE PRESS SYNDICATE OF THE UNIVERSITY OF CAMBRIDGE
The Pitt Building, Trumpington Street, Cambridge, United Kingdom

CAMBRIDGE UNIVERSITY PRESS
The Edinburgh Building, Cambridge CB2 2RU, UK
40 West 20th Street, New York, NY 10011–4211, USA
477 Williamstown Road, Port Melbourne, VIC 3207, Australia
Ruiz de Alarcón 13, 28014 Madrid, Spain
Dock House, The Waterfront, Cape Town 8001, South Africa

http://www.cambridge.org

© J. C. Naylor 2002

This book is in copyright. Subject to statutory exception
and to the provisions of relevant collective licensing agreements,
no reproduction of any part may take place without
the written permission of Cambridge University Press.

First published 2002

Printed in Italy at G. Canale & C. S.p.A.

Typeface Trump Medieval 9.5/13pt   *System* QuarkXPress™   [SE]

A catalogue record for this book is available from the British Library

ISBN 0 521 80925 8 hardback

All diagrams drawn by the author, John Naylor

Contents

Preface

This book is about things that can be seen in the sky. We all look at the sky from time to time, though usually it is to check the weather. By and large we don't look at it for enjoyment, in part because we don't know what to look for. Very few people who are unfamiliar with the many wonderful sights to be seen in the sky accidentally notice halos or sundogs, two of the most common optical phenomena. To be sure of seeing these and other sights, you must know what to look for and when to look. This is where I hope this book will come in useful. It has been written to help you find your way around the sky, and see for yourself the many wonderful things that it has to offer.

My earliest memory of looking at the sky is of having the three stars that make up Orion's Belt pointed out to me. I can't recall what I made of them; I remember being told that they are distant suns, though that didn't mean much to me at the time. I was, I think, six or seven years old.

It was, nevertheless, a defining moment, the start of a lifelong fascination with the sky. But for many years that interest was overwhelmingly bookish. I read about the stars, but I didn't look at them; or, at any rate, not often. And when I did, it was invariably a brief, careless, unreflective glance. I looked, but I didn't see. And I didn't see because I didn't really know what to look for.

The turning point was a book by Marcel Minnaert, a Dutch astronomer. The book was *The Nature of Light and Colour in the Open Air*, and it made me look afresh at things that I had all but ignored, and look out for things which I had never seen.

For the first time in my life, I saw one of the most common sights in the sky, an ice halo around the Sun; I noticed a mirage of a distant island; I spotted a *heiligenschein*, the faint glow sometimes visible around the shadow of your head; I counted the colours in a rainbow. I began to take notice of the atmosphere itself, and of how it alters and transforms the way things appear to us. I was amazed at how much there was to see. All I had to do was keep my eyes peeled, something I could do while looking out of the window, pottering about in the garden, or walking to work.

It wasn't long before I found myself searching the night sky for similar sights. I began to realise that the night sky is about much more than stars and planets. It is a dynamic entity. Gradually I fell in step with its rhythms: the nightly procession of stars, the monthly race around the heavens between the Sun and Moon, the yearly coming and going of planets, and much else.

As I learned more about the sky, the idea of this book took shape. I have called it *Out of the Blue* because its deals with phenomena that most of us have only seen by accident in the sky. It is a guide to a vast range of optical and astronomical phenomena, many of which occur daily, and which can be seen and understood without instruments or specialised knowledge. It has been written for anyone who is curious about, puzzled by, or just downright ignorant of the many optical phenomena that can be seen in the course of daily life. It offers practical advice about where and when you can expect to see these things, what you will see, and how to improve your chances of seeing them. It gives equal weight to the night sky and the day sky, and deals only with phenomena that can be seen with the naked eye. This is, after all, how most of us see things: we don't usually have telescopes or binoculars to hand.

I also hope that the book will help you make sense of what you see. I have assumed that you are a casual observer, and don't have a single focus of interest. You enjoy nature but are not, for example, a dedicated amateur astronomer or meteorologist. The text is thus a mixture of description and explanation, and draws on science, history, literature, mythology and anecdote. Explanations assume little or no scientific knowledge. Technical terms are kept to a minimum, are explained in the main text, and again in a glossary at the end of the book. Should you want to look into a particular phenomenon in greater detail, there is a comprehensive reference section in the sources and notes.

I could not have written this book without a great deal of help from a good many people. I should like to thank Jos Widdershoven, a one-time student of Marcel Minnaert, for introducing me to the sky; Gerald King for asking a lot of awkward questions; Alastair McBeith for reading an early draft, and setting me right on astronomical matters; Pekka Parviainen for his generosity and patience, not to mention his stunning photographs; Claudia Hinz and Francisco Diego similarly; several anonymous reviewers who spotted mistakes and made helpful suggestions; Mairi Sutherland for her hard work knocking my prose into shape and helping me to express myself more clearly; Faith Evans, my agent for dealing with the business side of things; and, most importantly, Sue, my wife, who never doubted that this book would see the light of day, for her encouragement, constructive criticism and much, much more.

Introduction

The art of seeing nature is a thing almost as much to be
acquired as the art of reading the Egyptian hieroglyphs.
John Constable, quoted in C.R. Leslie, *Memoirs of
the Life of John Constable*, Phaidon, 1951, p. 327

The first great mistake that people make in the matter, is the
supposition that they must see a thing if it be before their eyes.
John Ruskin, *Modern Painters*, vol. 1,
George Allen & Sons, 1908, p. 54

However extraordinary it may seem, it remains a fact that the
things one notices are the things with which one is familiar; it
is very difficult to see new things even when they are before our
very eyes.
Marcel Minnaert, *The Nature of Light and
Colour in the Open Air*, Dover, 1954, p. vi

Since this book is intended to encourage you to use your eyes, unaided by a
telescope, I should begin by saying a few words about the eye. Many people
assume that the eye is like a camera, and that therefore we must see whatever is before our eyes. But, to see, you must look. Seeing is a conscious act
and, unless you look actively, and have some idea of what you are looking
for, you probably won't see many of the things mentioned in the subsequent
pages. The eye is not a camera; it is an extension of the brain, the organ of
thought. Observation thus favours the prepared mind. One of the most
important lessons for anyone interested in what the sky has to offer is not
to take seeing for granted. It is something that you have to work at, and
which involves knowledge and patience in almost equal measure.

You should also not underestimate the power of the eye. It is widely
believed that looking at the sky without a telescope, particularly at night, is
a waste of time. Compared with the telescope the unaided eye seems a feeble
thing, hardly up to the challenge of seeing the faint, the small, or the unusual.

But the eye is a Jack-of-all-trades, a superbly flexible organ which can scan the entire sky in a matter of seconds, and take in with a single glance the spectacle of a clear dark sky, the majesty of the Milky Way, or the thrill of a meteor shower. These are not things that can be done with a telescope. And the more you use your eyes, the better you become at spotting things.

The other great advantage of relying on your eyes rather than on a telescope is that your eyes are always with you. Most astronomical telescopes are not very portable, and you won't always have one to hand when there's something worth seeing. If you rely on only your eyes, you'll never be caught short when there is something to see. This is not to say that a telescope isn't worth having. Far from it: the Moon's surface, or the disc of a planet, seen through a powerful telescope, are extraordinary sights. But a telescope does not, in itself, overcome the major disadvantage most people face when looking at the sky: not knowing what to look at, or how to look at it.

While on the subject of optical instruments, there is the important question: to photograph, or not to photograph? I have found that photography can get in the way of looking carefully, and enjoying a sight. In any case, you soon discover that many of the sights that you want to photograph are quite common. If, like me, you are not a particularly skilled photographer, and lack the patience necessary to improve your skills, leave photography to others. I now seldom carry a camera. I believe that the only point of lugging a camera around is for those once-in-a-blue-Moon events, and of which there are few, if any, good images. There are occasions when I could have kicked myself for not having a camera to hand, like the time I saw Manhattan looming above the horizon some 40 km across water from the shores of Connecticut. But I did have binoculars, and I spent the time exploring the sight, rather than fiddling around with the controls of a camera. In fact, I would say that, given a choice, binoculars are more useful to a skywatcher than a camera. The decision, of course, is yours. There are some grand sights to be photographed, as you can tell from the photographs in this book.

What of U.F.O.s, or unidentified flying objects? Leaving aside the issue of whether aliens have visited the Earth from another world, a subject too vast and contentious to be dealt with here, there will be occasions when you will see something in the sky that perplexes you. Don't jump to conclusions; there will almost certainly be a rational explanation. In defence of my scepticism, I offer my own U.F.O. experience. One night some years ago, in the city where I live, I saw four or five glowing ovals flying swiftly overhead. I had hardly glimpsed them out of the corner of my eye than they vanished behind some nearby buildings. They moved silently, one behind the other, weaving slightly from side to side. I realised that they were not aircraft or I would have heard their engines. What were they? For years after, I remained perplexed by what I had seen. Then one night, by chance, I saw a similar formation. This time I had binoculars to hand. I quickly raised them to my eyes, focused, and saw geese. As soon as I did so I could see why they looked like fast-moving glowing ovals. In the first place they were flying directly over me just above roof height so they took only a few seconds to cross my entire field of view, which made their apparent speed very great indeed. Secondly, they made no noise, something that had added to my confusion the first time I saw them. Finally, the glowing ovals were their wings, which reflected the orange glow of the streetlights below. This has been my only U.F.O. sighting to date. I have no doubt I shall see other things that puzzle me, and which I may be unable to explain, though I don't expect ever to see a flying saucer.

Finally, a few words about the rewards of skywatching. Sadly, light pollution due to street lamps prevents us from seeing all but the brightest stars. Astronomers complain that today's skies, especially city skies, are barren wastelands, devoid of all but a handful of stars, and hardly worth a second glance. But although light pollution makes it all but impossible to see faint stars and galaxies, including our own Milky Way, it doesn't prevent us from seeing the Moon, planets and the brightest stars. In fact, where the Moon and planets are concerned, you can see almost as many of the Moon's many aspects from most cities as you can from the darkest countryside. And, at their brightest, planets can be seen without difficulty even from the centre of a large city.

Above all there is the pleasure of simply allowing your eyes to roam until they fasten on something interesting, and of recognising it for what it is. The chances of seeing something hitherto unknown, or extremely rare, are slim. If you are new to skywatching, you will make many discoveries, but they will be of things that are new to you. If you are unfamiliar with the sky, you soon find that nature is brimming with optical phenomena that you have never heard of, let alone seen. If you keep your eyes peeled, there is a chance that you will see something that is new to you almost daily, or see something familiar in a new light. When you first see such phenomena you experience a thrill of discovery – always a magical moment. But it won't take you long to realise that most of these sights are everyday events, and that it was merely ignorance that stopped you noticing them sooner.

Gradually your experience of nature acquires another dimension as you realise that what we see is the result of the passage of light through the atmosphere, of reflection and refraction in the world around us. Rainbows and halos are particularly striking examples of this, but they are not in a special category. They are explicable by the same optical principles as ordinary shadows. When you understand this, experience stops being a disconnected series of seemingly one-off events. You begin to think of the sky as a single entity, and you find that you can anticipate phenomena.

The greatest thrill comes when you are sufficiently familiar with the sky to know when something out of the ordinary is happening. But the greatest reward is not in seeing something unusual, but in realising that it is unusual. This serendipitous state cannot be achieved overnight. You have to acquaint yourself with the everyday before you are in a position to notice the unusual. This takes time, patience, and a certain amount of commitment, though it's never a chore. In fact, that is part of the fun.

Chapter 1 | DAYLIGHT

> It is a strange thing how little in general people know about the sky. It is the part of creation in which nature has done more for the sake of pleasing man, more for the sole and evident purpose of talking to him and teaching him, than any other of her works, and it is just that part in which we least attend to her
>
> John Ruskin: *Modern Painters*, vol. 1, section III, ch.1 'Of the Open Sky'

1.1 The colour of the daytime sky

Our atmosphere is a thin, transparent layer of air, not much more than 100 km deep, that separates us from the dark and starry void beyond. On cloudless nights we look out at this fathomless darkness, and only the twinkling of stars betrays the presence of the atmosphere through which we gaze. But when the Sun is up, the whole sky glows brightly, and the dark abyss is concealed from us by an almost tangible blue dome that seems forever out of reach, beyond even the highest clouds and the furthest horizon.

There is no blue dome, of course. Instead during daylight hours we are immersed in what is known as airlight, which is the glow of the atmosphere when it is illuminated by the Sun. We look *through* airlight rather than at it, though it is so faint that we notice it only when looking through several kilometres of atmosphere. A blue sky seems distant because of the cumulative effect of airlight from points near and far. Most airlight, however, reaches us from the atmosphere within a few kilometres of where we happen to be. Although there are traces of air 100 km and more above the Earth's surface, the density of the atmosphere diminishes rapidly with altitude, so much so that 75% of its mass lies within 10 km of sea level. To give you some idea of what this means, on a 25 cm globe, this depth of atmosphere is equivalent to the thickness of a sheet of paper. The remaining 25% of air that lies more than 10 km above ground becomes ever more tenuous until it finally peters out some 150 km above our heads.

Sadly, Ruskin's remarks about people's knowledge of the sky are as true today as they were in his day. Ask almost anyone what the colour of the sky is, and they will almost certainly reply that it is blue. But, on closer inspection, everyone can see that, although the dominant colour of a clear sky is blue, it is never uniformly blue. It varies from one part of the sky to another,

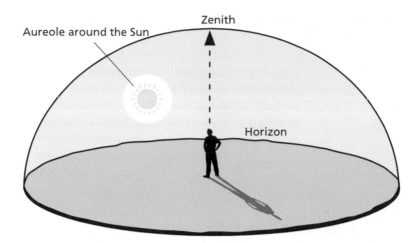

Figure 1.1 During daylight a cloudless sky appears to be a flattened dome. The zenith is the point in the sky directly above your head; it seems closer than the horizon, which is the circle where the sky appears to meet the Earth. When the Sun is close to the horizon, the darkest, bluest part of the sky lies at the zenith. The sky is less blue towards the horizon. Around the Sun there is often a whitish aureole.

and from day to day. Nor is it as blue as it could be: the vivid, saturated blue visible in the spectrum that is formed when a narrow beam of sunlight passes through a prism. Don't take my word for any of this. Step outside on any clear day, and look around the sky for a few minutes. Looking at the sky for yourself is far more enlightening, and much more fun, than simply reading about it. The things to concentrate on are its colour and brightness.

Begin by looking for the bluest part of the sky. Where is it in relation to the Sun? Compare its colour and brightness with that of the horizon. With a small mirror you can examine both at the same time: while looking at the horizon hold the mirror level with your eye, and adjust it so that you can see a reflection of another part of the sky. In this way you can directly compare any two parts of the sky.

Now shield your eyes with an outstretched hand, and look towards the Sun, making sure that your hand covers the Sun at all times. Your eyes can easily be permanently damaged by gazing at the Sun for more than a few moments. What colour is the sky directly around the Sun? How bright is it compared with the rest of the sky?

If you have a panoramic view you may be able to notice the distance at which the colour of the sky becomes apparent. Compare the appearance of things that are near you with those that are far away, and look for the blue airlight scattered by the intervening air between you and the more distant objects. At what point does this become noticeable? A few hundred metres? Several kilometres?

If you have time, see how the Sun's height above the horizon affects the distribution of colour and brightness over the whole sky. Compare the sky at midday, mid-afternoon and sunset.

Finally, what of the sky's colour on different days? The difference between a really clear sky and one that is even slightly hazy is quite startling.

If you want to notice subtle variations in sky colour you'll find a cyanometer useful. This simple device was invented some 200 years ago by a Swiss physicist, Horace de Saussure. It consists of several numbered strips of card each painted a different shade of blue. Saussure's original cyanometer consisted of 16 strips. Shades of blue, ranging from dark blue to light blue, are created by mixing Prussian Blue with white.

In use, you simply hold up the strips and sort through them until you find one with a hue that matches that of the portion of sky that you are looking at, and make a note of the number. You will find that doing this helps you notice variations in sky colour in a way that cannot be done merely by passive observation.

1.2 Why is the sky blue?

'Why is the sky blue?' is one of those deceptively simple questions that children ask. As with so many naïve questions, there is no simple answer. In the first place, the question presupposes that the sky has a uniform colour, and that colour is its only important feature. If you have looked at the sky attentively you will know that on a clear day its colour and brightness vary from zenith to horizon. When the Sun is low, the sky is bluest and least bright at the zenith. As the Sun climbs higher, the zenith gradually ceases to be the bluest part. Near the horizon and around the Sun, the sky is brighter and less blue. In fact, at the horizon the sky varies from light grey to white, and is never blue. The colourless glow around the Sun, known as an aureole, is not present in exceptionally clean air such as that found in polar regions.

It's difficult to say when people first began to wonder about the colour of the sky. In part they would have been hampered by their lack of knowledge of the relationship between colour and light. Until the end of the seventeenth century, when Newton established that sunlight is composed of several colours, and that an object takes its colour from the light that illuminates it, the view was that colour is an inherent property of an object and that sunlight merely serves to illuminate what is already there. Nevertheless, some 200 years before Newton, Leonardo da Vinci drew attention to the fact that a piece of blue glass looks bluer if it is thicker. He suggested that if the inherent colour of the atmosphere was blue then the sky should look bluer near the horizon. Because this is not the case, he concluded that sky colours were due to some other cause. He favoured particles of moisture in the atmosphere. Newton himself believed that the sky's colour had the same cause as the colours that can be seen in a thin film of oil. These colours are . due to the destructive and constructive interference of light.

The first person to study the colour of the sky systematically was Horace de Saussure. He invented the cyanometer to help him keep a record of changes in sky colour from day to day and in different localities. Saussure believed that the atmosphere had no colour of its own, and that sky colours were due to vapours.

The correct explanation for sky colours finally emerged towards the end of the nineteenth century as a consequence of experiments made by an Irish physicist, John Tyndall. Tyndall carried out an extensive investigation of the circumstances in which bluish colours occur when a beam of white light passes through a medium composed of tiny particles.

We associate colour with pigments. A pigment is any substance that alters the appearance of a beam of light by absorbing some colours and reflecting others. Paints are pigments. But air is not pigmented. In this sense it really is almost colourless. All the colours we see in the sky during daylight, from the blue of midday to the rosy hues of sunset, are due to selective scattering by molecules of the gases that make up the atmosphere.

In a vacuum, a beam of light is invisible unless it enters the eye directly. But if this beam encounters a tiny particle, a fraction of its light spreads, or is scattered, in all directions. If the beam is bright enough, and the medium through which it passes contains a high concentration of small particles, we see a little of its light without having to look directly at the source. If the particles are very small, as molecules are, the degree of scattering depends on the wavelength of the light: the shorter the wavelength, the greater the degree of scattering. This form of scattering is known as selective scattering. This is what Tyndall was investigating, though in his day it was called Tyndall scattering of light.

The correct explanation for selective scattering was given by an English physicist, John Strutt, at the end of the nineteenth century. (Strutt inherited the title of Lord Rayleigh from his father so selective scattering is also known as Rayleigh scattering.) He showed that selective scattering occurs only when light encounters particles that are much smaller than the wavelength of light. A typical molecule of gas is over a thousand times smaller than the wavelength of light, which is why selective scattering occurs within the atmosphere.

Strutt also showed that the degree to which the shortest wavelengths of light are scattered by very small particles is some 10 times greater than it is for the longest ones. This is why the colour of scattered sunlight is heavily weighted in favour of the bluer end of the spectrum. We perceive the shortest wavelengths as violet and blue and the longest ones as red. In fact about 40% of the light from the bluest part of a clear sky is composed of shorter wavelengths. The presence of longer wavelengths in scattered sunlight makes the sky appear less blue than it would if only the shorter wavelengths were scattered. Incidentally, although violet has a shorter wavelength than blue, the sky doesn't look violet because there is more blue than violet in sunlight. Furthermore, our eyes are more sensitive to blue than they are to violet, so violet light appears less bright than blue light of the same intensity.

Molecules are very weak scatterers of light, which means that the amount of light scattered by each molecule is tiny. So, despite the vast number of molecules that make up the atmosphere, the total amount of scattered sunlight within the atmosphere is small compared with the total amount of light that reaches the Earth from the Sun. Scattered light appears far brighter and more vivid than it really is because we see it against the dark void beyond the atmosphere. The whole sky right down to the horizon appears bright because almost every molecule scatters some light towards the ground.

The diminishing blueness towards the horizon is due to multiple scattering, scattered light that is itself scattered when it encounters further molecules. Light from the zenith encounters fewer molecules on its way through the atmosphere to the ground than light from the horizon does because the amount of atmosphere in this direction is several times less than it is in the direction of the horizon. Zenith light therefore looks bluer because it retains a larger proportion of shorter wavelengths compared with longer wavelengths than light from the horizon does. On the other hand, light scattered from the furthest horizon encounters many more molecules, and so blue light from the horizon is scattered away from the observer's line of sight to a greater degree than scattered red light.

You might expect that this would make distant objects appear redder than they in fact are. The reason the horizon looks white under normal circumstances is that scattered blue light reaches us from the air that lies between us and the horizon. The combination of long wavelengths from the distant horizon and short ones from the near horizon causes the horizon to

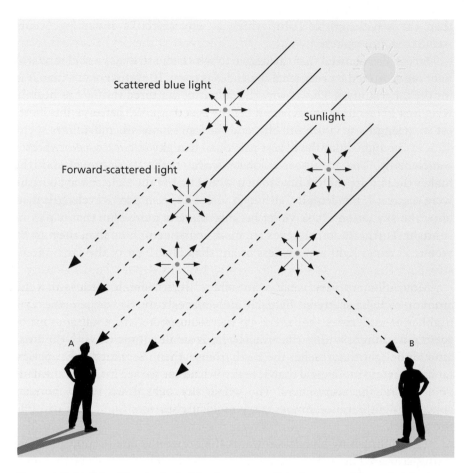

Figure 1.2 Airlight. The sky looks blue because sunlight is scattered by molecules in the atmosphere. A molecule scatters about ten times as much blue light as it does red light. Consequently, scattered sunlight, called airlight, contains far more blue light than unscattered sunlight. At the same time, forward-scattered light is deficient in blue light. In other words the proportion of red light to blue light is greater than in unscattered sunlight. From a direction perpendicular to the Sun's rays (B in diagram) airlight looks blue, whereas looking towards the Sun (A in the diagram) the sky looks less blue.

look white. The preponderance of long wavelengths over short ones in light from the distant horizon is very obvious during a total eclipse of the Sun, when the horizon turns orange. During the totality, when you are within the Moon's shadow, the atmosphere immediately around you is not directly illuminated by the Sun, and so you don't receive enough scattered blue light to make the distant rosy horizon look white.

The greater amount of atmosphere in the direction of the horizon also explains why the horizon is brighter than the zenith. Far more light is scattered from the horizon than from any other part of the sky because there are many more molecules in this direction than there are in any other direction. Hence, even in the absence of the sub-microscopic particles responsible for haze, the atmosphere is always brightest in the direction of the horizon. Conversely, the least bright part of the sky is at the zenith because the amount of the atmosphere in this direction is less than in any other direction.

Seen from a plane, or the summit of a high mountain, the zenith is noticeably bluer, though much less bright, than it is from sea level. It becomes less bright the higher you are, until you reach the edge of the atmosphere, at which point the sky is completely dark. In fact, the best place to see the variation in the colour and brightness of a clear sky is from a plane. Next time you fly look out of the window and compare the horizon to the zenith.

Both the colour and brightness of a clear sky depend on the density of the atmosphere. If the Earth's atmosphere were denser than it is it would contain many more molecules per unit volume. In these circumstances even light from the zenith would be multiply scattered. Within such an atmosphere the zenith would resemble the horizon: i.e., it would be both bright and white. If the atmosphere were less dense, then the zenith might be dark and colourless, and the horizon blue. Neither of these alternatives sounds very appealing, and one can't help feeling that our present atmosphere is a pleasing compromise between colour and brightness.

1.3 Airlight

Light scattered within the atmosphere is known as airlight to distinguish it from sunlight, which is light that reaches us directly from the Sun. The molecules of gas that make up the atmosphere are major source of airlight. Airlight due to molecules is noticeably blue, which is why the dominant colour of the atmosphere is blue. However, molecules of gas are not the only natural source of scattered blue light. Even particles that are many times larger than a molecule of gas scatter selectively, and are sources of blue airlight when they are present in the atmosphere. The distinctive blue hazes that are characteristic of many regions covered by vegetation are due to such particles. The origin of these particles is probably organic vapours exuded by plants. Chemical reactions between these vapours and ozone create scattering particles of the right size, i.e. approximately $\frac{1}{10}$ the wavelength of light. The Blue Mountains in New South Wales, Australia, and the

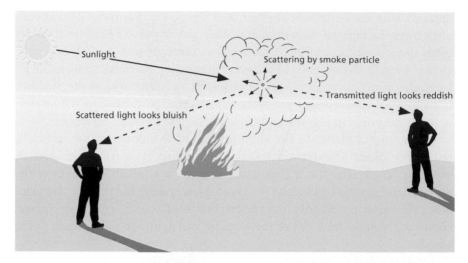

Figure 1.3 The colour of smoke sometimes depends on the direction from which it is viewed. The scattered light looks bluish and the transmitted light looks reddish.

Blue Ridges of Tennessee, USA, to name but two, are regions famed for their blue hazes.

Particles larger than $\frac{1}{10}$ the wavelength of light are another source of scattered light in the atmosphere. Unlike molecules, they scatter all wavelengths equally, though they do so mainly in the forward direction. Larger particles also scatter more light than smaller ones. This is why a haze composed of large particles tends to be bright and white, particularly when seen in the direction of the Sun.

Large particles suspended in the atmosphere are called aerosols. In an aerosol-free atmosphere, i.e. one composed entirely of molecules of gas, a so-called 'molecular atmosphere', the sky around the Sun would be blue. The fact that the Sun is almost always surrounded by a bright, colourless aureole is evidence that aerosols are invariably present in the atmosphere. In polar regions, where the atmosphere contains far lower concentrations of aerosols than it does at lower latitudes, the sky is blue almost up to the edge of the Sun. In fact a simple test for a clean atmosphere is to cover the Sun's disc with a finger held at arm's length, and compare the sky immediately around the Sun with that at some distance from it. In a really clear sky, the two areas should be indistinguishable because aerosol-free skies are blue skies.

Particles ranging in size from 5 to 50 times the wavelength of light are responsible for so-called grey hazes. There are many sources of such particles: dust, soot, salts, and acid droplets produced by chemical reactions between ozone and either sulphur dioxide or nitrogen dioxide. Many salts

Figure 1.4 The sky during daylight. The blueness of a clear sky is least at the horizon and increases towards the zenith. (*Photo* John Naylor)

and acids are hygroscopic, which means that they have an affinity for water. This makes them ideal nuclei on which water vapour can condense even when the air is not very humid. In these circumstances water vapour condenses on the hygroscopic nuclei, and forms tiny droplets. The resulting haze is sometimes called a 'solution-drop' haze. This type of haze will disperse if humidity decreases although, since the nuclei are still present, the haze will re-appear if humidity increases again. A hazy morning sky will often clear later in the day as the air warms up and humidity decreases.

The particles responsible for blue vegetation hazes are also hygroscopic which is why the best examples of these hazes occur in semi-arid regions.

In more humid environments vegetation hazes tend to be less blue because condensation increases particle size. The Blue Mountains of New South Wales are sometimes cloaked in a colourless haze, rather than a blue one, for precisely this reason.

Smoke from a fire is often black or brown. The colour is due to absorption of light by large particles of carbon. There are, however, many sources of smoke in which the particles are small enough to scatter light selectively. Take a look at smoke from bonfires, smouldering cigarettes, or the exhaust of a car that is burning too much oil. When strongly illuminated and seen against a dark background this smoke takes on a bluish hue, whereas when seen against a bright one it looks distinctly brown. A dark background highlights any blue light scattered by the particles of smoke. On the other hand, the shorter wavelengths of light from a bright background are scattered as they pass through the smoke. Incidentally, cigarette smoke will look blue only if it is not inhaled. Exhaled smoke looks white because smoke particles grow larger when in the lungs, either because they stick together or because water condenses on them. These larger particles are not able to scatter light selectively.

In rare circumstances, airlight can be reddish. This is usually due to fine dust, which is blown into the air in large amounts, say during a dust storm or a volcanic eruption. You can find more about reddish airlight in section 9.11 on Blue Moons.

1.4 Aerial perspective

We are all familiar with the way in which distant features in a scene are usually washed out compared with those that are near. We expect that extensive views should fade into the distance. If you spend time outdoors you will know the degree to which scenes vanish into the distance varies from day to day, and on some days the more distant parts of a scene are lost from view altogether. The name given to this phenomenon is aerial perspective.

Aerial perspective is due to light scattered by molecules and particles in the atmosphere. Scattering affects the appearance of distant objects in two ways. In the first place, some sunlight is scattered in your direction by the atmosphere that lies between you and the object. This scattered light is, of course, airlight. Secondly, some light reflected by the object does not reach you because it is scattered away from your line of sight, reducing the brightness of the object. Both processes are at work at the same time, and their

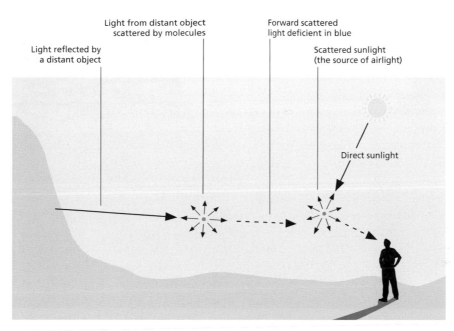

Figure 1.5 Aerial perspective. Light reflected by a distant object becomes gradually less blue and less bright as it makes its way towards us because of scattering by molecules that make up the atmosphere between the object and the observer. The reason that distant objects appear bluish is that we see them through a veil of airlight due to direct sunlight scattered by the atmosphere that lies between the object and observer.

combined effect reduces the contrast between the object and its surroundings. The result is that we seem to be looking at it through a bright veil. Given a high concentration of scattering particles, or sufficient distance between object and observer, contrast can be reduced to the point where the object vanishes from view altogether. This is why there is a limit to how far we can see through the atmosphere.

Aerial perspective is most apparent when you look out over gently undulating countryside. The sudden change in brightness between successive ridges makes the increasing brightness of airlight with distance much more obvious than the gradual change in tone seen over flat countryside. Aerial perspective is more easily noticed if there are objects in the foreground because they frame the view and supply the contrast necessary to notice the veil of airlight between you and objects in the middle distance.

An absence of aerial perspective can be disconcerting since we unconsciously rely on it to judge how far things are from us. When the air is very clear, distant objects that are normally seen through a veil of haze come into

Figure 1.6 Airlight. The increasing blueness of a distant landscape is due to airlight. (*Photo* John Naylor)

sharp relief, giving a scene an air of unreality. In the absence of aerial perspective most people find it difficult to estimate how far away distant objects really are, particularly in unfamiliar situations.

Mountaineers often encounter this problem because air at altitude contains few particles compared with air at sea level. Astronauts on the airless Moon found it particularly difficult to judge distance since, not only was there no airlight, but they found themselves in completely unfamiliar terrain.

Charles Darwin had this to say about the effect of an absence of aerial perspective on mountaineers:

> The increased brilliancy of the Moon and stars at this elevation, owing to the perfect transparency of the atmosphere, was very remarkable. Travellers having observed the difficulty of judging heights and distances amidst lofty mountains have generally attributed it to the absence of objects of comparison. It appears to me, that it is fully as much owing to the transparency of the air confounding objects at different distances, and likewise partly to the novelty of an unusual degree of fatigue arising from little exertion, – habit being thus opposed to the evidence of the senses. I am sure that this extreme clearness of air gives a peculiar

Figure 1.7 Aerial perspective. Airlight causes a distant hill to appear washed out compared with those close to us. (*Photo* John Naylor)

character to the landscape, all objects appearing to be brought into nearly one plane, as in a drawing or panorama.

Charles Darwin

The rosy hue of clouds, snow-capped mountains and other distant whitish surfaces at sunrise or sunset is due to sunlight that is deficient in short wavelengths. But aerial perspective can make distant clouds and snow look distinctly yellowish or even salmon pink at other times of the day.

Rosy hues are also seen at the horizon when the sky is almost completely overcast except for a narrow strip of sky along the horizon. The distant horizon will be in direct sunlight, but the air between the horizon and the observer will be shielded from the Sun by cloud. The same thing occurs during a total eclipse, when the horizon turns orange because the atmosphere around you is within the Moon's shadow.

The reverse effect is observed when the atmosphere is clear: distant objects take on a bluish hue. This is due to molecular airlight, airlight due to the molecules of gas of which the atmosphere is composed, which becomes noticeable when seen against a dark background. Mountains 'become blue when they become air', wrote John Ruskin.

Airlight stands out well against a distant tree-covered slope because each tree offers an irregular surface, some of which will be illuminated directly by sunlight, and some of which will be in shadow. Shadows provide an ideal background against which to see airlight. In these surroundings, molecular airlight may receive a boost from the blue haze characteristic of vegetation.

1.5 How far can you see?

The song has it that on a clear day you can see forever. But it's not true. Even on the clearest day, aerial perspective due to molecular airlight places a limit on how far we can see. On most days aerial perspective is increased by haze, which is why few of us have any real idea of what good visibility is like.

Visibility or, to give it its proper name, visual range can be a matter of life and death. Think how often poor visibility leads to accidents. Visibility also greatly affects our enjoyment of a scene: a bright haze that prevents us seeing distant objects, or which reduces them to mere outlines, renders any landscape unappealing and oppressive. Almost any scene will give you a lift when the air is free of haze, and everything is seen in vivid colours and sharp relief.

I suspect most of us take visibility as we find it. Only when we meet extremes – say when enveloped by fog, or when the atmosphere is particularly transparent – are we likely to be aware, despite ourselves, of the enormous difference that the state of the atmosphere makes to a view.

Given the choice, I imagine everyone prefers a scene that is not veiled in haze, which is to say that we all enjoy the consequences of good visibility. At the other end of the scale, many people are quite taken by fog, which lends an air of mystery to the most humdrum surroundings. No one likes haze. Haze is horrible: not the blue haze of airlight, but a veil of bright haze that washes detail out of a scene.

On a mundane level, visibility depends on our ability to see things clearly enough to identify them. To identify an object it must, at the very least, stand out against its background, if only as a colourless shape. The minimum condition for visibility is therefore a matter of relative brightness (or contrast), and not of colour or detail (e.g. individual features like windows and doors in a house).

Aerial perspective reduces the contrast between an object and its surroundings. Airlight acts like a bright veil between object and observer. At the same time, some light from the object fails to reach us, reducing its apparent brightness. Together these processes can reduce the contrast between an object and its surroundings to the point where they are indistinguishable, and the object vanishes from view.

Even the molecular airlight from a perfectly clear atmosphere can reduce the contrast between an object and its surroundings below the threshold of visibility. Since scattering depends on wavelength, visual range is greatest for red light, and least for blue light. Calculations based on the minimum amount of contrast that enables one to distinguish an object from its background yield a maximum visual range in perfectly clear air of 330 km in green light.

The atmosphere, however, is never free of aerosols that scatter a much greater amount of light than molecules do. Airlight due to these aerosols is very much brighter than that due to a purely molecular atmosphere, which is why aerosols have such a huge effect upon visibility. Compare the visibility for various atmospheric conditions (i.e. different particle sizes and concentrations) given in the table of visual range. Visibility in a light haze is some 10 times less than it is in what is taken to be an exceptionally clear atmosphere. Note that, from the point of view of visibility, a fog is merely a haze made up of particularly large particles, which scatter a much greater amount of light than smaller particles.

Paradoxically, visibility is often better when the sky is completely overcast than it is on cloudless days. Extensive cloud cover reduces airlight because it prevents the Sun from directly illuminating the atmosphere below the clouds. But it has no effect on the degree to which the atmosphere scatters light coming from the object, since this depends only upon the size and concentration of particles in the atmosphere. This suggests that airlight has a greater effect on visibility than other factors do.

Airlight prevents us from seeing stars during the day because it is brighter than the brightest star. After sunset, airlight diminishes gradually, and one by one stars become visible.

Visual range under different conditions

Condition	Maximum visual range
Dense fog	Less than 50 m
Thick fog	50 m to 200 m
Moderate fog	200 m to 500 m
Light fog	500 m to 1000 m
Thin fog	1 km to 2 km
Haze	2 km to 4 km
Light haze	4 km to 10 km
Clear	10 km to 20 km
Very clear	20 km to 50 km
Exceptionally clear	More than 50 km

Visibility can change from hour to hour, day to day, season to season, and from one location to another. There is evidence that visibility is actually getting less in some parts of the industrialised world. A major culprit is sulphur dioxide emitted by power stations that burn fossil fuels. Sulphur dioxide is a cause of an increase in solution-drop hazes, hazes that are formed when water condenses on tiny particles with an affinity for water. To escape this haze that threatens to envelop us permanently you will have to travel to remote spots, or climb to the tops of mountains that are at least 3 km high.

1.6 Polarised light

Polarised light is common in nature. Nevertheless, we are usually unaware of it because the human eye is unable to distinguish between light that is polarised and light that is not polarised without the use of artificial aids such as the polarising filters fitted to some types of sunglasses.

Polarisation is a property of transverse waves, which are waves that vibrate in a plane perpendicular to the direction in which they travel. A wave is said to be polarised if its vibrations are confined to a single plane. All electromagnetic waves are transverse waves, which is why light can be polarised.

A beam of light is made up of a huge number of individual waves. These don't all vibrate in the same plane as one another, and so the beam as a whole is unpolarised. Almost all sources of light, such as the Sun, electric lights and candles, are unpolarised. Their light can be polarised by removing waves that vibrate in planes other than a chosen one. This can be achieved if the unpolarised beam is reflected, refracted, scattered or absorbed. The resulting beam is said to be linearly polarised because most of the waves within the beam are vibrating in the same plane.

The simplest way to tell if light is polarised is to look at it through a polarising filter. If there is a change in brightness as you rotate the filter then the beam contains some polarised light. The change in intensity is due to the action of the material of which the filter is made. Only light that is polarised in a particular plane can pass through the filter. Light that is not polarised in this plane is partly or wholly absorbed. The filter has only to be rotated through a quarter turn to extinguish light that is polarised.

(a)

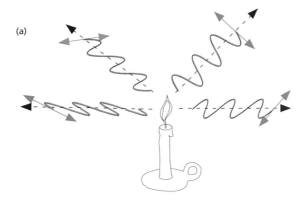

(b)

Unpolarised light   Polarising filter   Polarised light

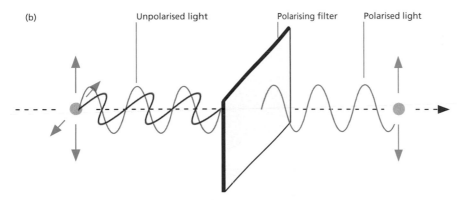

Figure 1.8 Polarisation of light. (a) Each light wave emitted from the flame vibrates in a different plane from its neighbours. Hence light from the flame is, as a whole, unpolarised.

    (b) When an unpolarised beam passes through a polarising filter, only waves vibrating in a particular plane emerge. In these diagrams unpolarised light is represented by crossed arrows. Polarised light is represented by a single arrow.

1.7 Polarised light from the sky

Almost all airlight is polarised to some extent. If you explore both the distribution and the direction of polarisation of light in a clear sky using a polarising filter you will find that the greatest amount of polarised light comes from a direction that is perpendicular to the rays of light from the Sun. When the Sun is low in the sky, most of the polarised light comes from a band some 30° wide, which arches overhead in a direction that is perpendicular to the Sun's rays. Light from the horizon, and from the Sun's aureole, is hardly polarised at all (see figure 1.10).

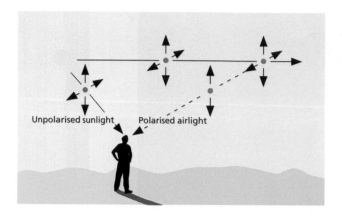

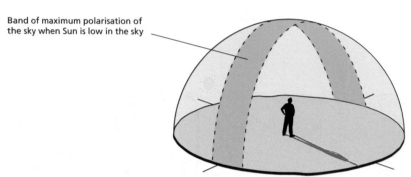

Figure 1.9 Polarised airlight. In this diagram, unpolarised light is represented by crossed arrows. This is because unpolarised light consists of electromagnetic waves that are vibrating in all possible planes perpendicular to the direction in which the light is travelling. Light that reaches the eye directly from the Sun is unpolarised. On the other hand, scattered sunlight is polarised. The polarisation is greatest in light that is scattered from sunlight perpendicularly to the direction in which the light travels. Hence the most highly polarised skylight comes from a broad band perpendicular to the Sun's rays.

Light from clouds is not polarised, and so they remain equally bright whatever the orientation of the filter. This is most noticeable when clouds lie within the region of maximum polarisation of the sky.

Another effect can be noticed when looking at distant clouds. A polarising filter not only reduces airlight from the sky beyond a cloud, it also reduces the amount of airlight scattered by the air between you and the cloud. If the air is free from aerosols, this airlight is bright and blue; reducing it allows you to see a distant cloud more clearly. Light reaching you from the cloud is, of course, deficient in short wavelengths, and so should have a rosy hue. Unfortunately a polarising filter absorbs light and thus reduces the

Figure 1.10 Polarised skylight. Without a polarising filter to block out polarised sky-light the sky appears as shown in the top photograph. The dark wedge in the centre of the sky in the bottom photograph is due to normally invisible polarised light that has been absorbed by the polarising filter. (*Photo* John Naylor)

Figure 1.11 Polarised skylight. If polarised airlight is filtered out with a polarising filter distant objects are seen more clearly. Compare the view of the Moon seen in the top picture without a polarising filter and the one in the bottom picture taken through a polarising filter. (*Photo* John Naylor)

brightness of colours seen through it. Nevertheless, there is a perceptible shift in hue, and this shows up very clearly in a photograph. You will notice the same effect when looking at the Moon in daylight. It is clearly shown in the pair of photographs on the opposite page.

1.8 Polarised light due to reflections

When light is reflected, some of it is absorbed by the reflecting surface. Smooth, non-metallic objects absorb light that is polarised in a direction perpendicular to the surface; the reflected beam is largely polarised parallel to the surface. In this context smooth means optically smooth, or smooth on the scale of the wavelength of light. This seems like a tall order, but, surprisingly, many surfaces satisfy this condition: a sheet of glass, the surface of a pond, and varnish on a piece of wood are all optically smooth. Consequently many reflections contain a significant proportion of polarised light.

Polarisation by reflection also depends on the angle of incidence of the light. The maximum polarising effect occurs at what is known as the Brewster angle. Light reflected at this angle is completely linearly polarised in a plane parallel to the reflecting surface. However, polarisation is not confined to this angle: some polarised light is reflected at all but the largest angles of incidence. For example, although the Brewster angle for glass is typically 57°, more than half of the light reflected from glass at angles of incidence between 30° and 80° is polarised. The fact that polarisation is not confined to a particular angle of reflection is handy: polarising sunglasses would be useless if light were polarised only in reflections at the Brewster angle. As it is, most reflections, whatever the angle, contain some polarised light, and polarising filters can be used cut out a noticeable proportion of this light.

In polarising sunglasses the filters are aligned so that only light that is polarised perpendicularly to the ground can pass through them. This alignment is chosen because this type of sunglasses is designed to be used in situations where it is desirable to reduce the amount of light reaching the eyes from horizontal surfaces. If you tilt your head when looking at a horizontal surface through such glasses, the surface will get brighter because horizontally polarised light can now pass through the filters.

Using a polarising filter to investigate your surroundings you will find that, generally speaking, reflections from smooth, dark surfaces contain a greater proportion of polarised light than reflections from surfaces that are rough and light coloured.

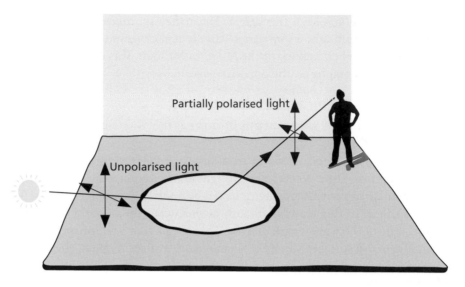

Figure 1.12 Polarisation by reflection. Sunlight reflected in water is polarised because light vibrating in a horizontal plane is more efficiently absorbed by the reflecting surface than light that is vibrating vertically. hence the reflected light contains a greater proportion of light that is vibrating vertically.

A rough surface is made up of many tiny smooth ones. Each of these reflects light polarised in a particular direction. These polarised reflections are themselves reflected several times by adjacent surfaces before emerging back into the air. Multiple reflections jumble up the polarised light, and consequently the surface as a whole does not give a polarised reflection. A light-coloured surface is bright because it reflects more light than one that is dark. The brightness of the unpolarised reflection overwhelms any polarised light reflected by the surface. The amount of polarised light in a reflection increases if the reflecting surface is wet because this makes it smoother.

A polarising filter reveals an altogether more colourful world since an object's true colour is often masked to some degree by surface reflections. Look at shiny surfaces such as the leaves of an evergreen plant, grass, water and polished or varnished wood, and note the change in appearance when the filter is rotated to reduce the shininess. Reflections from metallic objects are also polarised, though the degree of polarisation is not as much as it is in the case of non-metals and tends to be greatest for large angles of incidence.

All halos and rainbows are polarised. In the case of the rainbow the plane of polarisation of its light follows the curve of the arc. The reason for this is explained in section 5.20. Halos are such a diverse group of phenomena that each type needs to be considered separately. Both the 22° and the 46° halo are weakly polarised in a direction that is radial to the centre of the halo. Halo phenomena that are due to reflection, such as sun pillars and parhelic circles (see sections 7.8, 7.9), are more strongly polarised than refraction halos (see section 7.1). Interestingly, a mirage is not polarised, which might be taken to be proof that it is not caused by reflection, except for the fact, mentioned above, that light reflected at large angles of incidence from non-metallic surfaces is hardly polarised at all. Moonlight, being a reflection of sunlight, is weakly polarised though this is difficult to detect using ordinary polarising filters because the change in intensity that they bring about is not large enough to be noticeable to the eye.

1.9 Haidinger's brush

Although, strictly speaking, humans are blind to polarised light, it produces an effect in the eye known as Haidinger's brush. This is a faint, yellowish, oblong patch some 2° to 3° long in the centre of the field of vision that can be seen when looking at a broad expanse of polarised light. The orientation of the oblong is determined by the plane of polarisation of the light – its longest axis is at right angles to the plane of polarisation. Some people are more sensitive to this effect than others and can apparently see it simply by looking up at a clear blue sky. Alternatively, you can look at a white surface such as a sheet of paper or a cloud through a polarising filter. Images fade from view when stationary because the eye is rapidly fatigued, so Haidinger's brush is most easily noticed as the filter is rotated through a quarter turn, and the figure suddenly jumps 90 degrees in the opposite direction.

The presence of the figure in the middle of your field of view can be distracting when looking at other effects caused by polarisation. Nevertheless, since the longest axis of the figure is always at right angles to the plane of polarisation, you can use it to check the alignment of the polarising plane of the filter.

Many insects are sensitive to polarised light. Bees make use of the distribution of polarised light in a clear sky to navigate to and from their hive. There is, however, no truth in the claim that Vikings used naturally occurring polarising filters as a navigational aid. It is said that they navigated by

Figure 1.13 Haidinger's brush. This is a faint yellowish pattern visible when you look at light that is polarised. The orientation of the pattern depends on the plane of polarisation of the light, as shown in the diagram.

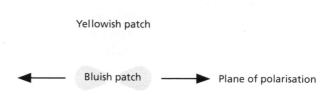

Yellowish patch

Bluish patch ⟶ Plane of polarisation

the Sun, and used such filters to determine the position of the Sun when the sky was overcast. Unfortunately for this theory, skylight from an overcast sky is not polarised. Furthermore, no one has ever found one of these polarising filters used by Vikings.

Chapter 2 | SHADOWS

> I recommend to those who are new to these games the
> entertainment of watching the gyrations and transformations of
> their own shadows while walking at night along a lamplit road.
> As you pass close to the lamp your shadow will appear short
> and squat by your side, and slowly turn in the direction of your
> walk while growing longer and narrower, till the bright lamp of
> the next lampost will replace it by the shadow that is now
> behind you.
>
> E.H. Gombrich: *Shadows: The Depiction of Cast Shadows in
> Western Art*, National Gallery Publications, 1995, p. 12

2.1 No light without shadow

A shadow is a volume of space not directly illuminated when light is intercepted by an object. Usually we are aware of these shafts of darkness only
when they fall upon an illuminated surface, where they are seen as dim, distorted outlines. But if the medium through which the shaft of the shadow
passes is filled with particles able to scatter light, such as dust-laden or hazy
air, fog, or turbid water, the shaft itself becomes visible.

We are apt to overlook the effect of shadows on the way the world looks
to us. If we notice them at all, it is probably as a diversion. We can amuse
ourselves by noticing how our shadow changes shape depending on where
we stand relative to the source of light, or on the contours of the surface on
which the shadow falls. And people have always been beguiled by the
dancing shadows produced by the flickering flames of a fire or a candle. But
since shadows are cast by objects, they are indirect evidence of space and
solidity. Adding shadows to the objects in a painting makes them seem more
real, and lends depth to the scene in which they appear.

Although our eyes instinctively seek out light, without the subtle contrasts between brightness and darkness brought about when light meets
solidity, the world would look flat and drab. Recall for a moment the monot-
ony of an overcast day, made all the more oppressive by an absence of
shadows. In fact, the pattern of light and shade created when a scene is illuminated is one of the best visual clues that we can have as to the shape of
things within it. It emphasises the relief of uneven surfaces, and provides

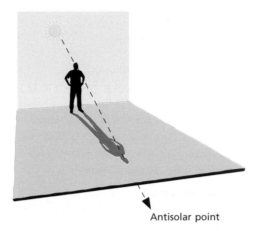

Antisolar point

Figure 2.1 The antisolar point. This is a point on the opposite side of the sky to the Sun. It lies in the direction of the shadow of your head.

clues to depth, enabling us to discern the shape and form of otherwise featureless snow-covered landscapes and clouds. Despite the fact that Galileo's telescope was not powerful enough to reveal precise details of lunar features, Galileo deduced that the Moon's surface was like the Earth's surface, mountainous and uneven, from the elongated lunar shadows that he saw through the telescope.

It hardly needs saying that all shadows point away from the source of light. In the case of solar shadows this direction is known as the antisolar point. This is an imaginary point that lies on the opposite side of the sky from the Sun, on a line that passes through the eye of the observer. This means that your antisolar point is in a different place from mine. We can't share the same antisolar point. The antisolar point is of interest because so many optical phenomena, such as rainbows, glories and *heiligenschein*, are centred on it.

2.2 Solar shadows

Leaving aside the Moon, which in any case shines by borrowed light, the Sun is our only source of natural light. Because it appears to us as a disc, rather than a point like a faraway star, the Sun illuminates everything with rays of light that are not quite parallel to one another. Such a source of light is known as an extended source. Close to an object the effect of this divergence is negligible, and so shadows are sharpest when they fall near an object. Further away the divergence increases, giving the shadow a fuzzy edge known as the penumbra; the darker, central portion is known as the umbra. Although the umbra receives no light directly from the Sun, it is illuminated by airlight, and by sunlight reflected from surrounding surfaces. Together these lighten solar shadows. The fuzziness of the penumbra is due to the fact that the amount of direct sunlight illuminating the penumbra increases towards its outer edge, making it difficult to discern the point at which the shadow gives way to full sunlight.

The penumbra is such an obvious feature of solar shadows that it comes as a surprise to learn that the first person to describe it, and explain how it arises, was Leonardo da Vinci in the fifteenth century. Earlier accounts of shadows used the fact that a shadow has a similar shape to the object that causes it to argue that this was possible only because light travels in straight lines. In other words, shadows were used as proof that light travels in straight lines. The penumbra was ignored because sunlight was represented as if it came from a single point on the Sun, rather than many points spread over its disc. Shadows formed by a point source lack a penumbra. However, Leonardo's analysis of shadows was not widely known, and the world had to wait another hundred years for Johannes Kepler to lay the foundations for the science of optics in its modern form, and in passing explain the penumbra. In fact, we owe the term 'penumbra' to him.

The angular diameter (see Appendix, page 301) of the Sun is approximately 0.5°, or $\frac{1}{120}$ radian, and so the maximum distance at which the umbral shadow can be formed is approximately 120 times the diameter of the object. Beyond this point, only the penumbral shadow persists. To cast an umbral shadow at a particular point, an object must have an angular diameter that is greater than that of the Sun measured from the point at which the shadow falls. Hence neither small objects that are close to a surface, nor large ones that are far from one, can cast an umbral shadow. It's worth keeping this in mind where eclipses and crepuscular rays are concerned. Eclipses, of course, are due to the grandest of all shadows: those that the Earth and Moon cast into the void. We are unaware of the Moon's shadow until we pass through it and witness an eclipse of the Sun. In an eclipse of the Moon, the Moon passes through the Earth's shadow.

The Moon is our only other major source of natural light. At its brightest, moonlight is about half a million times less bright than sunlight. How do shadows cast in moonlight compare with those cast in sunlight? The most striking difference between them is how profound lunar shadows appear to be compared with solar ones. In daylight, scattered and reflected

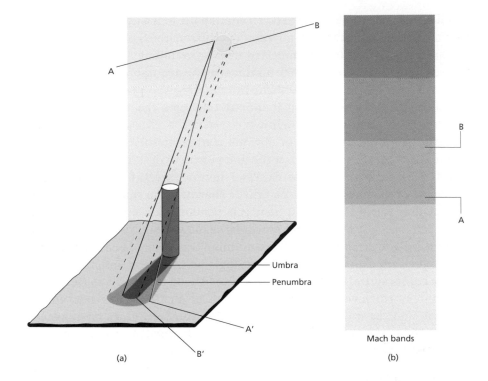

(a)   (b)

Figure 2.2 (a) Solar shadows. The Sun is not a point source of light and so shadows formed in sunlight have a fuzzy edge called the penumbra. The central region of the shadow receives no light directly from the Sun, and is called the umbra. In this diagram, the width of the penumbra has been greatly exaggerated to show its relationship to the umbra. In reality, it is much less broad. The penumbra is fuzzy because some light reaches it from the Sun's edge. For example, although no light from A reaches the area between A' and B', this area receives light from B.

(b) Mach bands. Where a darker rectangle meets a lighter one, the edge A of the darker one is noticeably darker than its opposite edge B.

sunlight are bright enough to illuminate all but the deepest recesses. At night, only those nooks and crannies that are directly illuminated by moonlight are visible, hence the inky shadows and dramatic contrasts that characterise a moonlit scene. At the same time, a lunar shadow appears sharper than a solar shadow because its penumbra is not very bright. Nevertheless, even in moonlight, the branches at the top of a tree cast shadows with quite obvious penumbras. To see this clearly, catch the shadow on a sheet of white paper.

Shadows play an important role in the way the Moon appears to us, both to the naked eye and through a telescope. Compared with solar shadows on Earth, those on the Moon are very much darker because of the absence of an atmosphere that would otherwise scatter light into the shadow. There *is* scattered light on the Moon: sunlight reflected by its surface. But much of this doesn't reach the depths of the pits and hollows that cover the Moon's surface. Solar shadows on the Moon are thus rather like lunar shadows on Earth: very dark. This is why the surface of the crescent Moon appears so much fainter than it does at other times during the cycle of lunar phases (see section 9.18 for a fuller explanation). At the same time, shadows cast by craters and mountains when the Moon is waxing or waning emphasise relief, and this makes the lunar landscape far more real when seen through a telescope than it would otherwise appear. It is this that enabled Galileo to conclude that the lunar surface is uneven, not smooth as his contemporaries believed.

A great deal of light reaches the ground, even on days when the sky is completely overcast. Despite this, distinct shadows are not formed because sunlight is scattered as it passes through clouds, so that it is fairly evenly distributed across the whole dome of the sky when it emerges from them. Under these conditions there is as much light coming from one direction of the sky as from any other direction and so a shadow can't be cast. Nevertheless, if the sky is partially blocked off, say by a wall, then a feeble shadow is cast in the direction of the obstacle.

2.3 Shadows formed by point sources

The shadow cast by an object when it is illuminated by a point source differs in several important respects from one that is cast in direct sunlight. To begin with it has no penumbra.

A point source is one that is very much smaller than the object it illuminates. A street lamp or a candle flame both act as point sources for objects that are at some distance from them. The difference between such sources and the Sun is that rays of light from the lamp or candle effectively diverge from a single point, whereas those from the Sun come from several points spread across the solar disc. Consequently, the edge of a shadow cast by a point source is always sharp no matter how far from an object it is formed. The size of the resulting shadow depends on how close the object is to the

Figure 2.3 Shadow due to a point source. When the source of light is a point, or almost so, shadows cast are sharp (they don't have a penumbra) and increase in size with distance from the object, so that even a small body can cast a huge shadow.

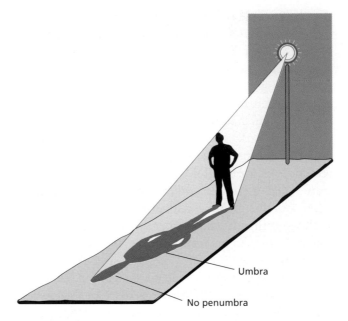

Umbra

No penumbra

source of light, and on the distance from the object to the surface on which the shadow is formed.

Stars and planets are point sources, though their light is usually too feeble to cast shadows. Nevertheless, at its brightest, Venus casts umbral shadows that are just noticeable to a dark-adapted eye outdoors on clear moonless nights far from city lights. Even Jupiter and Sirius are bright enough to cast noticeable shadows if light from all other sources is rigorously excluded, say by viewing the shadow within a darkened room. Light from the planet or star enters the room through an open window, and the shadow of a suitably positioned object is cast onto a white surface. Your eyes must, of course, be fully dark adapted to notice the very faint change in brightness that marks the boundary of the shadow. A shadow cast by Sirius is more difficult to see than one cast by Jupiter.

On occasion the Sun can act as a point source. Look carefully at a shadow when the Sun's disc is about to be obscured by a sharp-edged cloud: just before the Sun completely disappears behind the cloud, shadows become much sharper because the Sun has been reduced to a point source, if only for an instant. The same thing occurs shortly before and after totality during a solar eclipse: shadows become very sharp.

Figure 2.4 Shadows due to a point source. This photograph was taken shortly before totality. The Sun was a narrow sliver rather than a disc, and resulting shadows lacked a penumbra. (*Photo* John Naylor)

2.4 Mach bands

Try the following experiment. Fasten a small coin to a windowpane that faces the Sun. Hold a sheet of white card close to the window so that the shadow of the coin falls on it. Slowly move the card away from the window, keeping an eye on the shadow as you do so. You will find that, when the card is approximately 100 coin-diameters from the coin, the umbral shadow appears as a dark ring surrounding a noticeably less dark centre. At the same time the area just beyond the penumbral shadow appears to be brighter than the rest of the card. The same pattern of apparent increased darkness and brightness is seen if the card is held so as to catch the shadow of the window frame or indeed of any object at sufficient distance from the card.

With a little practice, these variations in brightness can be noticed out of doors around almost any shadow, though success depends on the colour and texture of the surface on which the shadow falls. For example, examine the edge of the shadow of your upper torso and head cast on a cement pavement. If the Sun is low in the sky and therefore you cast a long shadow, you should be able to notice that the more distant parts appear to be fringed by a narrow, bright band. This bright band falls just outside the penumbra. Now shift your attention to the umbra: it appears darkest just inside the outer edge.

In all these situations there is no physical reason for what you see. It is a purely physiological phenomenon due to the fact that, when a light-sensitive cell within the eye is stimulated, the response of neighbouring cells to the same stimulus is lessened. This is an evolutionary adaptation that makes it easier for our visual system to detect shapes by emphasising changes in brightness that occur at their edges. The physiological process is known as lateral inhibition. An unwanted consequence of lateral inhibition is that, when you look at the boundary between areas of unequal brightness, the eye's response makes the brighter side of the boundary appear lighter than it really is, while simultaneously making the darker side look less bright. The phenomenon was first systematically investigated by Ernst Mach, an Austrian physicist and philosopher of distinction, and is known as a Mach band. A Mach band is present wherever there is a sudden change in the brightness of a surface. It can sometimes be confused with the after-image caused when your eyes stray from a bright area to a darker one. You will find that with a little practice you can easily distinguish the one from the other.

Although Mach bands are illusory, they can be photographed in the sense that the eye responds to variations in brightness in the photograph in exactly the same way as it does to the original. In other words, you will see a Mach band in a photograph of a Mach band, although neither has an objective existence.

2.5 Coloured shadows

An umbral shadow out of doors would be totally dark were it not that it is indirectly illuminated by airlight. On a clear day this is, of course, predom-

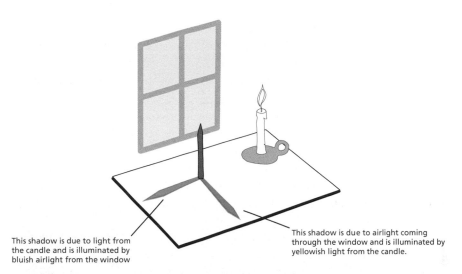

This shadow is due to light from the candle and is illuminated by bluish airlight from the window

This shadow is due to airlight coming through the window and is illuminated by yellowish light from the candle.

Figure 2.5 Coloured shadows. The shadow of an object may be coloured if it is illuminated by light from a different source than the one that produces the shadow. In this diagram a pencil casts two shadows. One is due to airlight coming through a window. This shadow is illuminated by candle light and looks yellowish. The shadow cast by candle light is illuminated by airlight and looks bluish.

inantly blue, and shadows sometimes have a noticeably blue hue. Shadows on snow are unmistakably blue. With practice, you should be able to notice a blue hue to shadows cast on surfaces such as cement pavements, which are not as white as snow.

Distant shadows take on a bluish hue because airlight scattered by the intervening atmosphere becomes more noticeable when seen against a dark background. The shadow acts as a backdrop against which we can more easily notice the faint blue hue of airlight.

Strongly coloured shadows are also produced if an object is simultaneously illuminated by more than one source. A shadow formed by the light from a north-facing window on a clear day will look distinctly yellow if the object that casts it is simultaneously illuminated by the light of a filament lamp: the shadow cast by the lamp will be blue. Coloured shadows can also be seen outdoors soon after sunset when objects are illuminated from one direction by blue airlight, and from another by a low-pressure sodium street-light. A shadow cast by the yellowish sodium light is distinctly blue. The colour can be so vivid that at first glance it may appear to be painted on the surface on which it is seen. Low-pressure sodium lights also cast coloured shadows in combination with moonlight or car headlamps.

The colours we ascribe to these shadows depend only partly on the

Figure 2.6 Blue shadows. A shadow cast on snow looks distinctly blue due to the fact that it is illuminated by blue airlight. (*Photo* John Naylor)

colour of the illuminating light. The eye itself plays a role. When we look at two adjacent surfaces, the colour of one is affected by that of the other. Painters have long known this. To see how this comes about, imagine that you are looking at a circular blue patch on a red surface. Better yet, cut out a circle of bright blue paper and glue it to a sheet of vivid red paper. When you shift your gaze from the red surface to the blue circle, your eye's blue-sensitive cells respond strongly to the blue surface because they were previously unstimulated when illuminated by light from the red surface. At the same time, the red-sensitive cells have been fatigued and so do not respond strongly to any red that might be reflected by the blue patch. The overall effect is to make the blue look bluer than it would if it were not surrounded by a red surface. If the whole surface were the same colour (say all blue or all red), the response of the cells sensitive to that colour would gradually weaken and the surface would look less vivid than it did to previously unstimulated cells, i.e. as it did when you first looked at it. We can generalise this as follows: a patch of a particular colour will appear more vivid if it is surrounded by another colour that does not stimulate the same type of colour-sensitive cells. The phenomenon is known as simultaneous colour contrast, and is similar to lateral inhibition.

Now consider a blue shadow seen out of doors at twilight just after the streetlights have been turned on. The area around the shadow is illuminated by the orange–yellow light of a sodium lamp. The area within the shadow is illuminated by scattered airlight which is predominantly, though not wholly, blue. Hence, when you look at the shadow, the reflected airlight stimulates the blue-sensitive cells much more strongly than those sensitive to yellow. The result is a shadow that looks distinctly blue. The opposite effect occurs when an object is simultaneously illuminated by light from a north-facing window and a light from a lamp. Light from the lamp is deficient in shorter wavelengths, those at the blue end of the spectrum, and so the shadow cast by airlight is illuminated by light that is deficient in blue. The eye's response to blue is lessened by airlight, and so when looking at this shadow it responds more strongly to the longer wavelengths than to the shorter ones, and the shadow takes on a smoky yellowish hue.

J.W. Goethe, the renowned German poet and dramatist, investigated the phenomenon of coloured shadows in some detail. He carried out many experiments during a lengthy series of investigations that he conducted into the nature of colour in the hope of discrediting the Newtonian view of colour. It was a lost cause, in part because Goethe didn't draw a clear distinction between the physical and physiological aspects of light. But he was a meticulous observer, and he recorded a huge number of fascinating colour-related phenomena that are contained in his *Farbenlehre*, the book he wrote on his theory of colours. Among these is the following description of coloured shadows.

> Let a short, lighted candle be placed at twilight on a sheet of white paper. Between it and the declining daylight let a pencil be placed upright, so that its shadow thrown by the candle may be lighted, but not overcome, by the weak daylight: the shadow will appear of the most beautiful blue.
>
> W. Goethe

2.6 The *heiligenschein*

You may sometimes have noticed a faint sheen, or increased brightness, around the shadow of your head when this falls on a grass lawn, particularly when the Sun is low, and you cast a long shadow. This sheen is known as a *heiligenschein*, a German word meaning 'holy glow'. It is much more obvious when you are in motion, and its brightness varies from one patch of lawn to another.

To see a *heiligenschein*, stand on a lawn in which the blades of grass are erect and a few centimetres long. Rock slowly from one foot to the other and

Figure 2.7 Self-shadowing. Each of the white circles casts a shadow represented here by a black circle. These shadows become visible away from your antisolar point. Around the antisolar shadows are hidden by the objects and so this region appears brighter. The effect of self-shadowing in this diagram is seen more clearly if you hold the diagram at arm's length.

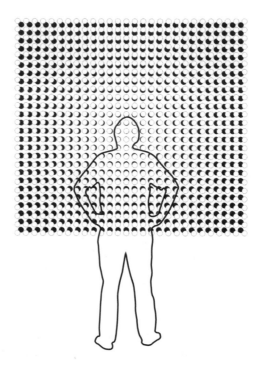

keep an eye on the brightness of the region just above your head's shadow. You should notice that, as the shadow moves, the grass above it becomes slightly brighter, while the grass above the point where the shadow fell previously is now noticeably less bright. Moving makes the *heiligenschein* more obvious because a change in brightness is more easily noticed than constant brightness.

A *heiligenschein* can also be seen from a considerable distance on fields of wheat and rice, both from ground level and from the air. In Japan the *heiligenschein* is known as *inada no goko* or 'halo in the rice fields'. It is frequently seen from balloons and aircraft when flying over woodland. Look for a *heiligenschein* around the antisolar point of a plane that you are travelling on, just after take off, or as it comes in to land. These are not the only circumstances in which you can see one, as the following report makes clear.

> . . . in this mountain region I have frequently seen the halo projected on a grassy slope a mile or more distant, and under these conditions it appears as a circular or elliptical patch of light without the central shadow, the diminution of intensity due to the penumbral shadow of one's head being, of course, quite inappreciable at such a distance.
>
> J. Evershed

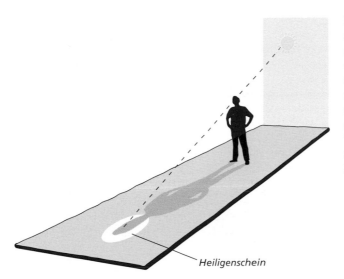

Figure 2.8 The *heiligen-schein*. A faint brightness is often seen around the shadow of your head when this falls on grass. It is most pronounced when you cast a long shadow, say in the early morning or late afternoon. The brightness is due to self-shadowing and is known as a *heiligenschein*.

Heiligenschein

Grass near the edge of the shadow of your head looks bright because a blade of grass at or near the antisolar point gets in the way of its own shadow. This is known as shadow hiding. Within this region, you see mainly the illuminated surfaces of blades, and few shadows. When you look away from the antisolar point, you see grass blades from the side, and so can see their shadows. The overall brightness of the grass thus diminishes when you shift your gaze away from the edge of your shadow. Given that shadow hiding is most effective at the antisolar point, it follows that you can't see a *heiligenschein* around someone else's shadow, and that they can't see yours.

The overall shape of the bright patch due to shadow hiding is determined by the general shape of the objects on which you cast your shadow (figure 2.8). Elongated objects such as blades of grass produce elongated bright patches. Hence the *heiligenschein* seen in grass is most pronounced just above the shadow of your head. In the case of a forest canopy seen from a plane or balloon, the *heiligenschein* is circular because the canopy of individual trees seen from above is also approximately circular.

Shadow hiding also plays a role in the variation of the Moon's apparent brightness as it orbits the Earth. The Moon's phases are due to an increase or decrease in the portion of its illuminated surface visible from the Earth. As the area of this visible surface increases from one day to the next, the Moon reflects more and more light in the Earth's direction.

Figure 2.9 *Heiligenschein*. The shadow of the photographer is surrounded by a bright glow due to light reflected back in his direction by dew drops on the grass. This glow is called a *heiligenschein*. The photograph can't show the *heiligenschein* around the shadow of his companion because light reflected in that direction doesn't enter the camera lens. (*Photo* John Naylor)

However, when the Moon is a crescent, this visible surface is illuminated by a low Sun, which leads to a great deal of self-shadowing. Self-shadowing makes the Moon's brightness less than it would be if it were a smooth sphere. As the Moon approaches opposition, the point along its orbit when it is on the opposite side of the Earth from the Sun, it gets closer to the anti-solar point, and shadows on its surface become shorter until, at opposition, they vanish from sight altogether, hidden behind the objects which cast them. In their place we see only the illuminated surfaces of these objects, i.e. shadow hiding takes place. This dramatic increase in the Moon's brightness is known as opposition brightening. Note that the Moon is not any brighter at opposition than it would be if it were perfectly smooth. The fact that it has a rough surface means that it is much less bright at phases other than full than it would be if it were smooth. Because Mars also has a rough surface, it is the only other astronomical body that undergoes noticeable opposition brightening. The other superior planets don't do so because they don't have solid surfaces and therefore have no shadow-casting features.

Self-shadowing and shadow hiding also play a part in the appearance of landscapes here on Earth. In the morning or in the afternoon, when the Sun is low in the sky, the sunward side of a rough surface is always noticeably darker than it is in the direction away from the Sun.

A *heiligenschein* seen on dry grass is sometimes referred to as a dry *heiligenschein*, to distinguish it from a *heiligenschein* seen on grass covered in dew. A dew *heiligenschein* is brighter than a dry *heiligenschein*.

If you examine dew on a blade of grass with a low-power microscope you will see that most dewdrops are small and spherical. This enables them to act as miniature lenses that focus sunlight on the surface of the leaf. The leaf reflects this light which then is channelled back more or less in the direction from which it came. Dew is deposited as spherical drops on some grasses more easily than on others. Hence dew *heiligenschein* are more pronounced on some lawns, or particular parts of a lawn, than on others.

A very distinct *heiligenschein* is formed on road signs that have been painted with reflective paint. This type of paint contains large quantities of tiny glass spheres to make it more reflective and these act as very efficient retro-reflectors. The signs are usually above head height. Your shadow can fall on such a surface only when you are illuminated by a light that is below head-height, say by the headlamps of a car. Occasionally you may see a very bright *heiligenschein* on street signs at sunset or sunrise. The Sun is then low enough for your shadow to fall on street signs that are close to the ground.

You can photograph your *heiligenschein*, but keep in mind that when you do so it will be formed around the shadow of the camera and so will only appear around your head if you hold the camera to your eye.

2.7 Shadows on water

Distinct shadows are cast on turbid water but not on clear water. On the other hand, reflections can be seen more clearly on water that is not turbid. The effects are related. If you examine shadows cast on turbid water you can see that they are formed within the water and not on the surface. The reason for this is that the suspended particles that make the water turbid also reflect a great deal of light back towards the surface. Those that fall

Figure 2.10 Shadows in water that is slightly turbid can be fringed with colour. Light from A appears reddish because some blue light has been scattered by the particles that make the water turbid. Light from B appears a smoky-blue because scattered light is seen against the dark background of the shadow.

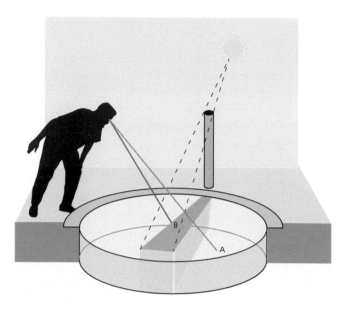

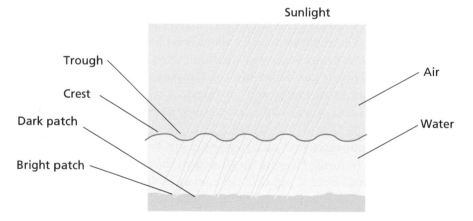

Figure 2.11 Patterns of light and shade in water. A crests acts like a converging lens, and concentrates light forming a small bright patch on the surface below. A trough has the opposite effect: it spreads light so that the surface below is faintly illuminated.

within the shadow are less brightly illuminated than those in direct sunlight and this provides the contrast which is necessary if you are to see a shadow.

On the other hand, the brightness of turbid water makes it difficult to see reflections in it. The best surface reflections are seen in pools of clear

Figure 2.12 Shadows in water. The streaks of light that stream out from the shadow of the hat in this photograph are due to shafts of light within the pool of water in which they were seen. (*Photo* John Naylor)

water with dark bottoms. The amount of light reflected in this case represents only some 5% of the incident light. The remainder passes into the water where some is absorbed by the material at the bottom of the pool and the rest is reflected back towards the surface. It is possible, however, to see reflections in the surface of turbid water where the surface is in shadow. In this situation, the particles within the water are not illuminated and so reflect no light to overwhelm the faint reflections from the surface.

You will sometimes see a faint orange fringe around the edge of a shadow cast in slightly turbid water. The colour is due to selective scattering by particles suspended within the water, which are small enough to scatter shorter wavelengths preferentially. You see the orange hue within the shadow formed on the surface of the water because no surface reflections are possible from the area covered by this shadow and this allows you to see light reflected from the bottom of the pool. This light is deficient in shorter wavelengths because of selective scattering by the suspended particles. At the same time the scattered short-wave light is seen through the shaft of the shadow within the water, which, consequently, will have a faint smoky-blue hue. You will see this blue hue more distinctly if you use a polarising filter to reduce surface reflections. The effect may be more noticeable in a particular pond or stream on one day but not on another because of changes in the size and concentration of particles suspended in the water.

A final effect to look out for is a faint pattern of darting rays that appears to radiate from the shadow of your head if this falls on slightly turbid water when its surface is ruffled by a light breeze. Although the rays appear to be on the surface, they are in fact shafts of light within the water caused by the focusing action of waves criss-crossing the surface. These shafts of light are parallel to one another, but because of perspective they converge on the anti-solar point and so they appear to radiate from the shadow of your head. Since the phenomenon is seen in slightly turbid water, your shadow will sport a faint orange fringe. If you can't see these rays when your shadow has a tell-tale orange fringe, you can produce the rays by briefly stirring the surface with a stick. The pattern is seen best in water that is at least 40 cm deep, against a dark background and with waves criss-crossing the surface in random directions.

2.8 Shadows formed by clouds

Clouds cast shadows into the air. These are visible as faint reductions in sky brightness that fan out from the edge of any cloud that lies directly between the Sun and an observer. The shadows are most noticeable in a clean atmosphere because airlight due to a haze may be bright enough to mask the contrast that renders them visible. They appear to fan out from the parent cloud because of perspective. In reality, their sides are almost parallel because they are formed by sunlight, the rays of which do not diverge greatly. Perspective causes the part of the shadow that is nearest to you to look larger than the one furthest from you.

Dramatic examples of this effect may sometimes be seen above the horizon when the Sun is about to rise or at which it has just set. These twilight shadows are known as crepuscular rays (section 5.5). At other times of the day, the alternating pattern of light and shade emanating from a cloud are variously known as 'sunbeams', 'the Sun drawing water' or 'the backstays of the Sun'.

Looking at these diverging shafts of light, it is tempting to conclude that the Sun acts as a point source, and that its light spreads out in all directions. In other words, it appears as if sunlight illuminates the Earth from a single

Figure 2.13 Cloud shadows. Clouds cast shadows that are visible as diverging shafts of light and dark in the sky. (*Photo* John Naylor)

Figure 2.14 Contrail shadows. Condensation plumes left by the engines of high-flying aircraft cast shadows that become visible if there is a thin layer of cloud between the plane and the ground. The shadow, which is closer to the ground than the contrail, will always lie further from the Sun than the contrail due to perspective. (*Photos* John Naylor)

point. Yet, when we look at the way in which light and shadow are distributed over a cloud, it is obvious that this cannot be the case. Ignoring the slight divergence of the Sun's rays due to the fact that it is an extended source, we may assume that the Sun's rays are more or less parallel to one another. Hence the Sun illuminates everything around us from the same direction. This can be confirmed by looking at cumulus clouds that lie at some distance to the Sun's disc. Note which parts are in shadow and which are directly illuminated. If the Sun were a point source, it would illuminate the sunward side of these clouds, whereas these may well be in shadow because, from the point of view of the clouds, the Sun is really behind them, rather than off to one side.

Contrails are those long plumes of cloud that sometimes form in the wake of aircraft. They are due to condensation of the water vapour emitted by aircraft engines when the aircraft is flying through cool or moist air. If there is a thin layer of cloud below the aircraft, the shadow of a contrail may be visible from the ground as a faint dark streak in the cloud layer. The contrail shadow is subject to the same perspective considerations as any other cloud shadow and so to someone on the ground it will appear to lie further from the Sun than the contrail. Very occasionally, when an aircraft is flying directly towards or away from the observer, the contrail shadow lies directly between the observer and the contrail. If the Sun is low in the sky, the contrail shadow will fall ahead of the contrail, and it will seem as if the plane is following the line of the shadow. Contrail shadows can also be seen from an aircraft. You can find out more about this in section 6.4, which deals with glories.

2.9 Shadows under trees in leaf

Look at the shadow of any tree in leaf, and you will notice that many of the patches of sunlight within its shadow are oval. This is not accidental. The gaps between overlapping leaves act as pinhole cameras and each produces a separate image of the Sun. These images are not perfectly sharp because the gaps between the leaves are not small enough to be true pinholes. Furthermore, although the Sun is circular, its image in this situation is usually oval because its light strikes the ground obliquely. Where the Sun is overhead, the patches of light are circular. During a partial solar eclipse the oval patterns become crescent-shaped, and during an annular eclipse (see section 10.2) they form rings. The rarity of these patterns makes it well worth while recording them on film.

Figure 2.15 Shadow of a tree during an eclipse. The usual ovals of light that are seen within the shadow of a tree become crescents during the partial phase of an eclipse of the Sun. (*Photo* John Naylor)

You will find that the size of the leaf affects the coarseness of the patches of light. Compare those formed under a tree that has small leaves with those formed under a tree with large leaves.

Chapter 3 | MIRAGES

> Quite early in spring, on any warm cloudless day, this water-mirage was visible . . . an appearance of lakelets or sheets of water looking as if ruffled by wind and shining like molten silver in the sun. The resemblance to water is increased when there are groves and buildings on the horizon, which look like dark blue islands or banks in the distance, while the cattle and horse feeding not far away from the spectator appear to be wading knee-or belly-deep in the brilliant water.
>
> W.H. Hudson: *Far away and long ago*, J.M. Dent & Sons, 1939, p. 57

3.1 Atmospheric refraction

When looking at something, we not unnaturally assume that what we see is exactly as we see it and precisely where we see it. In almost all situations in which we find ourselves this assumption is perfectly justified because we are close enough to the things that we see for their light to reach us by travelling in a straight line to our eyes through a homogeneous atmosphere.

Over larger distances, however, the atmosphere is not homogeneous because its density is not uniform. As you might expect, the density of air decreases with height because the lower layers are compressed by the weight of those above. This, in turn, affects the speed of light through the atmosphere. Light slows slightly as air density increases, and as it slows a beam of light changes direction, or refracts. Consequently, light from a distant object may reach us from a direction that does not coincide with its true position. When this happens, the object will be seen displaced from where it really is. You will have noticed how refraction can displace the position of an object when looking at a teaspoon that is partially immersed in water. The spoon appears bent at the point where it meets the surface of the water. The effect of refraction in this case is very pronounced because refraction from water to air is far greater that it ever is within the atmosphere.

Small though it is, the cumulative effect of atmospheric refraction can extend our horizon beyond the point at which a straight line drawn from our eye just grazes the Earth's surface. Or, to put it less confusingly, refraction within the atmosphere enables us to see objects that are slightly below the true horizon. This condition is known as looming. Though you are probably unaware of it, a common example of looming occurs when the Sun is seen on the horizon. In fact, it is then physically just below the true horizon. The Sun, of course, is outside the atmosphere, so when we see it on the horizon its light has traversed the longest distance possible through the atmosphere, which is why atmospheric refraction has such a marked effect on how and where we see the Sun when it is on the horizon. The Sun's image appears above the horizon because as its light nears the ground it refracts so that its path is always concave to the ground (see figure 4.12).

For objects on the Earth's surface, the effect of atmospheric refraction is usually less marked unless there are pronounced temperature gradients within the air just above the ground, i.e. when air temperature either increases or decreases rapidly with height. Temperature affects the density of the air, hot air being less dense that cold air.

Although air temperature normally decreases with height, there are times when the ground is cooler than the air directly above it over a large area, so that air temperature increases with height, and the density of air diminishes more rapidly than normal. In these circumstances it may be possible to see objects that are usually well out of sight beyond the horizon, an instance of looming. Temperature gradients within the atmosphere can also cause distant objects to appear as if they are vertically stretched (towering), or compressed (stooping). A good example of stooping is the flattened appearance of the Sun or the Moon when they are on the horizon.

Temperature gradients are also responsible for mirages. What distinguishes a mirage from looming, is that an object seen in a mirage is accompanied by at least one inverted image of itself.

The term 'mirage' was coined by Gaspard Monge, an eminent French mathematician and ardent Bonapartist. Monge went to Egypt in 1798 as a scientist with Napoleon's expeditionary force. There he saw, and was much taken by, the striking desert mirages that were frequently noticed as the French army made its way across the almost featureless terrain between Alexandria and Cairo. On his return to France, he invented a word for this phenomenon from *se mirer*, which means, 'to be reflected', and lectured widely on the subject.

Of course, Monge was by no means the first person either to see or to describe such sights. He was well aware that sailors had long been familiar with these things. 'Looming' is an old nautical term. However, in Monge's day, scientists didn't distinguish between looming and mirages, lumping them all together as examples of what they called 'abnormal refraction'.

Mirages are associated in popular imagination with hot deserts. Yet they are common in cool climates too. The main reason that mirages are often overlooked is ignorance. Most people don't see them because they are

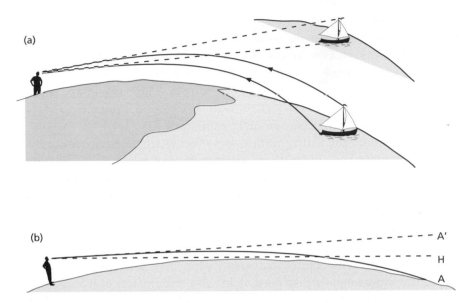

Figure 3.1 (a) Looming occurs when the air closest to the ground is much cooler than that above. Under these conditions light from a distant object is more strongly refracted towards the ground than normal. As a consequence it may be possible to see well beyond the true horizon. In this diagram we see that the rays from the hull of the boat are refracted slightly more strongly than those from the mast, which causes the object to appear squashed (or stooped) to the observer.

(b) The lower diagram shows the path of a ray of light through the atmosphere close to the ground. Light from a point A that lies somewhere beyond the true horizon H follows a curved path to the eye and appears to come from a point A′ that is above the true horizon. Since in these circumstances we see light from a point below the true horizon, the visible horizon itself will be more distant than under normal conditions.

unfamiliar with the phenomenon. Even when a mirage is seen, it is often dismissed as a trick of the eye, an illusion. But a perception is an illusion only if you believe that it doesn't have a physical cause. A mirage can be called an illusion only if you believe that what you see really is a pool of water rather than a mirage of the sky. In any case, only the most pronounced effects are usually noticed because objects in a mirage are often several hundred metres or more from the observer, and so the apparent size of things seen in a mirage may be half a degree or less. The unaided eye cannot easily resolve details of an unfamiliar object that has such a small apparent size and so nothing out of the ordinary is noticed, unless you know the tell-tale signs. To see mirages you should equip yourself with binoculars. Standard 7×35 binoculars are perfectly adequate (see Appendix for more details).

When looking at a mirage, keep in mind that the nature and degree of change in an object's appearance is determined principally by the temperature profile of the layers of air nearest the ground. The position of the observer is also important. The closer your eye is to the top of this layer the more pronounced is the mirage that you will see.

Though there are potentially a large number of possible temperature profiles, in practice you are only likely to encounter two types. If the ground is warmer than the air above it, you may see what is known as an inferior mirage. If the ground is cold and there is a layer of warm air a few metres above it, then you may see a superior mirage. Furthermore, depending on the rate at which temperature changes with height, superior mirages can take several different forms.

3.2 Inferior mirages

Inferior mirages are extremely common. Their most striking feature is a resemblance to a shimmering patch of water in the middle distance, but, unlike real water, these disappear when approached. They can be seen by anyone walking or driving along a straight road on a hot day. They are particularly obvious as you approach the top of a gentle incline ahead of which stretches a straight, horizontal section of road. Your eye is then placed just above the layer of hot air next to the surface of the road. Objects that happen to be in the right position will appear to be partially immersed in a pool of water.

The resemblance of this type of mirage to water is so compelling that it is the basis on which the phenomenon is named in many non-European languages: *nigemidu* ('escaping water') and *kagenuma* ('false swamp') in Japanese, and *shuiying* ('water image') in Chinese. There is, however, more to the inferior mirage than a semblance of water, though to see this clearly you have to use binoculars. Objects in a mirage are usually a long way from the observer, anything from tens of metres to several kilometres, which makes it difficult to see the details of a mirage clearly.

Ideally you should begin by looking at mirages of objects that are familiar to you: telegraph poles, motor cars, sailing boats, people, etc. Knowing what they should look like, you are in a better position to notice any changes in their appearance. You may, for example, just be able to notice that they are slightly drawn out or vertically elongated. This is towering. The most important changes, however, occur with increasing distance between object and observer where a point is reached when two images are seen: one erect

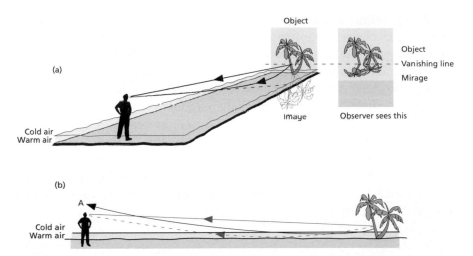

Figure 3.2 Inferior mirage. This type of mirage is brought about when the layer of air closest to the ground is much warmer than that slightly above it. Light from a distant object passing through the warm air is refracted upwards as shown in diagram (a). An observer therefore sees the object indirectly, because these refracted rays appear to him to come from a point below the object from which they come. This is the mirage. At the same time he sees the object directly through the cooler air. Hence he sees the object and what seems to be its reflection. He can't see all of the object directly because rays from points on the object closest to the ground are so strongly refracted that the pass over his head (for example ray A in diagram b). The point on the object below which he can't receive rays is known as the vanishing line.

and the other inverted. The inverted image is often compressed, known as stooping, and appears to be reflected in a pool of water. This pool is a mirage of the sky beyond the object. At greater distances an object may vanish from view, seemingly totally immersed in water.

Once you know what to look for, inferior mirages can be seen over almost any surface that is warmer than the air above it. Apart from roads and deserts, they can be seen on runways, beaches and meadows. They can also be seen over water – for example, lakes, bays, straits and the open sea. When looking out over water in mirage conditions, coastlines, islands and ships on the horizon all take on a characteristic appearance: their upper parts appear to float a little way above the horizon, which is itself slightly below the true horizon, while their lower parts are replaced by an inverted, compressed image of the upper parts.

An inferior mirage can also be seen by pressing the side of your face

Figure 3.3 Inferior mirage over water. The strip of sky that lies between the distant lighthouse and the land on which it is sited is an inferior mirage of the sky. Within this an inverted, miraged image of the lighthouse can be made out. (*Photo* Pekka Parviainen)

against a long flat brick or stone wall that has been warmed by the Sun to form a layer of very warm air next to the wall.

Mirages are usually associated with heat. Nevertheless, the principal factor that gives rise to a mirage is the rate at which air temperature decreases with height, not the actual temperature. Contrary to expectation, therefore, inferior mirages can, and do, occur above cold surfaces. They are common in cool climes where they may be seen over large expanses of water or ice when these are warmer than the air above, something that is more likely to be the case in winter than in summer.

An inferior mirage is due to the refraction of light as it passes through the air between object and observer. This refraction is caused by a change in air density which, in the case of mirages, is due to differences in temperature: the warmer the air the less dense it is. Hence light travels most rapidly through the warmest layer, refracting so that its path is concave to the layer of the coolest air.

The necessary temperature gradient is brought about when a surface is much warmer than the air above it. The layer of air in contact with the surface is heated, principally by convection, and becomes very much warmer than the air slightly further away from the surface. This creates a

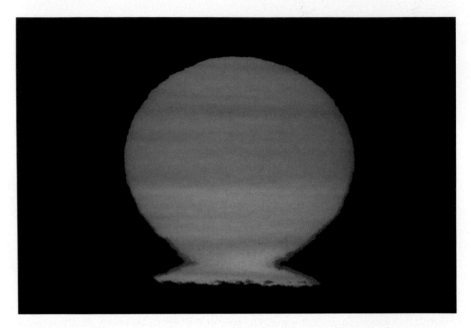

Figure 3.4 Inferior mirage of the Sun. The Sun is on the point of setting. The uppermost section of its disc is seen as an inferior mirage between the lower limb of the Sun and the horizon. The combined appearance looks like the Greek letter Ω (omega). (*Photo* Pekka Parviainen)

temperature gradient that is steepest within the layer closest to the surface. Since the differences in density that this brings about are very, very small, only rays that pass through this layer at very shallow angles (almost parallel to the ground) will be refracted so that they enter the eye from a direction that is slightly below the horizontal. This is why the source of these rays must lie at a considerable distance from the observer if it is to give rise to an inferior mirage, or the observer must have his eye very close to the ground.

Every point on an object sends light in many directions. The eye, however, can intercept only a narrow cone of rays from each of these points. Under normal circumstances, i.e. a small temperature gradient, we see objects by the light that they send directly towards us. This gives rise to a single image of the object. But under mirage conditions, some of the light that would otherwise not reach your eye is refracted as it passes through the layer of warm air next to the ground so that it enters your eye from below the point on the object from which it came. As can be seen in Figure 3.2, these rays appear to come from a point on the object below their true origin and give rise to a second image that seems to be a reflection of the upper portion of the object. This image is the inferior mirage. The line below which you cannot see the object directly is called the vanishing line. Everything seen below the vanishing line is inverted. If your eye is too high, it will not intercept the rays that produce the inverted image, and you will not see a mirage. Crouching down so as to place your eye just above the layer of warm air next to the ground brings the mirage into view.

The pool of water in which the objects in a mirage often appear to be partially immersed is an inferior mirage of the sky due to refracted airlight from beyond those objects. The similarity to a pool of real water is not accidental. The colour of the surface of a body of water under a clear sky is often a reflection of the sky's colour. Gaspard Monge was particularly impressed by the extensive mirages that made the desert appear as if it were flooded. He noted that under such circumstances the horizon appears to be noticeably closer than it really is: the whole world seems to shrink. This is to be expected because light from distant objects will be refracted so that they pass over the observer's head so that he sees only light from objects that are much closer to him.

Inferior mirages may have been responsible for a number of optical curiosities reported in antiquity. Several writers, including Monge, have suggested that, in the desert, entire armies have been, or can be, rendered invisible to their enemies through 'immersion' in an inferior mirage of the sky. Even more intriguing is the suggestion that the Biblical story of the parting of the Red Sea may have involved an inferior mirage. The event could be explained by assuming that the Israelites saw an extensive inferior mirage of the sky that they took to be a large body of water. As they approached, the 'water' appeared to recede: hence the parting of the waters. Later, when they looked back at the pursuing Egyptians, they saw them in a similar mirage and assumed that the Egyptians were engulfed by the same body of 'water' through which the Israelites had just passed.

An inferior mirage can be photographed. As a guide to the lens that you should use to obtain a satisfactory photograph of a mirage, think of the size of the image produced by your binoculars. If these are, say, 7 by 35 and have given a satisfactory image of a mirage then you will need to use a lens that gives 7 times the magnification of the standard lens, i.e. use a 300 mm to 400 mm telephoto lens with a 35 mm camera.

3.3 Superior mirages

When you see an inferior mirage of an object, at least part of the object is directly visible to you. This, of course, is only possible if it lies somewhere

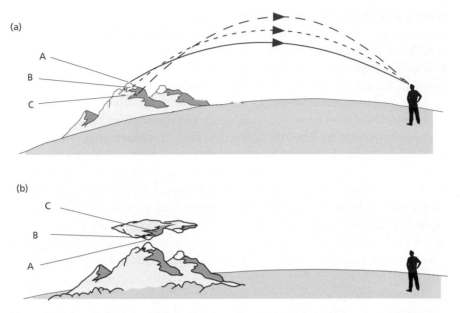

Figure 3.5 Superior mirage. If there is a temperature inversion a few metres or more above the ground an inverted image of a distant object may be seen. This is due to rays that are refracted towards the ground as they pass through the temperature inversion. Under such circumstances some rays reach the observer from the sky and in reverse order (diagram a) so that what is seen is a squashed, inverted image of the topmost part of the distant object. At the same time the object itself appears raised above the horizon and drawn out vertically (diagram b).

between you and the true horizon. If, however, there is a layer of air a few metres above ground in which temperature increases with height (instead of the normal situation in which it decreases with height), in other words a *temperature inversion*, then you may see images of objects that are beyond the horizon. This is the cause of a superior mirage.

A temperature inversion brings objects into view above the true horizon because rays of light are refracted within the inversion so that their path is concave to the warmest layers of air.

An observer thus sees light that would otherwise pass overhead. This light will appear to come from a point that is above its actual source, which is why, when there is a widespread temperature inversion, objects are seen above their true position. Since temperature gradients within the atmosphere are never uniform, the amount of refraction is not the same for all rays and this can lead to towering and stooping.

Figure 3.6 Superior mirage of land. A superior mirage of a distant archipelago appears to hover, inverted, above the archipelago. (*Photo* Pekka Parviainen)

There are several ways in which a temperature inversion can be brought about. For example, the lowest layer of a mass of air resting on a much cooler surface is cooled by the surface. At mid latitudes, such inversions frequently occur over lakes and bays on still, warm afternoons in spring and early summer. In Arctic and Antarctic regions, the source of warm air is usually a cell of high-pressure air sandwiched between weather fronts, while the cold surface is either a vast ice field, or a huge stretch of icy-cold water. In fact the optical effects of the superior mirage are sometimes referred to as high-pressure refraction. High-pressure air generally brings with it clear, stable conditions followed by short-lived but violent cold-front storms, and a feature associated with the Arctic mirage is excellent visibility followed by storms. Sometimes the temperature inversion occurs at some distance above the Earth's surface. These so-called 'lifted' inversions are responsible for unusually great visual range and the inverted images that characterise a superior mirage.

Temperature gradients within a lifted inversion can vary greatly with height: for example, the gradient may be more pronounced (i.e. there is greater change in temperature with change in height) a few metres above the ground than at ground level. In such a situation, the rays of light that reach

Figure 3.7 Superior mirage of a ship. The superior mirage of the ship appears as a mirror-image above the ship. (*Photo* Pekka Parviainen)

the observer through the layer of air closest to the ground will not be refracted as strongly as those that pass through the air at the level of the inversion. Under these conditions, the appearance of a distant object may undergo dramatic transformations because its image may tower or stoop strongly, and more than one image of it may be seen. In its most characteristic form, a superior mirage gives rise to three distinct images of an object: one on the horizon, with a reflected pair of images suspended some distance above it. The lower half of this pair is inverted while the upper half is compressed (i.e. stooped). Several examples of this type of superior mirage were seen by S. Vince when he looked out across the Straits of Dover from Ramsgate in 1798 'on August the first, from about half an hour after four o'clock in the afternoon till between seven and eight. The day had been extremely hot, and the evening was very sultry; the sky was clear, with a few flying clouds.'

What he saw were mirages of sailing ships, and of the coast of France. Although he used a telescope with a magnification of 30 to 40 times, he noted that the multiple images of the ships were visible to the naked eye. William Scoresby, a pioneer of Arctic exploration, saw similar mirages though a telescope when he was sailing in the icy sea east of Greenland, 'consisting chiefly of images of ships in the air.' during an evening of a beautifully clear summer day in 1820 during which there 'was scarcely a breath of wind. The sea was as smooth as a mirror.'

Don't imagine that superior mirages are to be seen only rarely, or only over the sea. Suitable conditions are frequently met at mid latitudes over lakes, bays and straits during the afternoon in late spring or early summer, giving us the chance to see superior mirages without having to journey to polar regions. Superior mirages can also seen over land if there is a suitable temperature inversion at heights ranging from tens to hundreds of metres from the ground.

3.4 Lake monsters

There is a good deal of evidence that abnormal refraction is the cause of the many apparent sightings of lake and sea monsters that have been reported from around the world. Take the example of the most famous and elusive of them all, the Loch Ness monster. Almost all reported sightings of 'Nessie' appear to have been made under ideal superior mirage conditions: calm, cold water overlaid by warmer air. Furthermore, several of the descriptions of this creature given by the more experienced monster hunters fit very well with what you would be likely to see under such conditions. In the first place, as you would expect if the object that is the source of the mirage were different each time, descriptions of the monster's appearance vary from one sighting to another. Secondly, the monster is said to move through the water noiselessly, and to appear and disappear without disturbing the surface.

Another source of mirage-generated creatures is to be found in Norse legends of sea monsters such as the merman. The Norsemen's description of the merman as an elongated figure with a large bulbous head and no arms fits in very well with the appearance of a triple-image superior mirage of either the snout of a killer whale or the head of a walrus rising up above the surface of the sea. Norsemen associated the sighting of the merman with danger. This seems a reasonable connection to make since, as has already been mentioned, the high-pressure weather that favours the superior mirage often ends abruptly with a storm when the cold front moves into the area formerly occupied by the mass of high-pressure air.

Given that there is a link between atmospheric conditions and strange sights, monster hunts might meet with more success if those involved in them were to keep an eye on the weather map, and embark on an expedition only when a high-pressure cell is imminent.

3.5 Looming and unusual visual range

The annals of Arctic and Antarctic exploration are full of accounts of unusually great visual range. There are good reasons for this. In the first place, polar air is much clearer than air at lower latitudes, so there is less airlight between the observer and the distant object, and less light from the object is scattered and absorbed on its way to the observer. Secondly, as we have seen, widespread temperature inversions often occur at high latitudes, leading to pronounced looming. The greatest visual range occurs when the inversion is some hundreds of metres above the Earth's surface, a 'lifted' inversion. These inversions also occur at lower latitudes and are frequently met with at sea.

An notable example of looming occurred over the English Channel in the summer of 1798.

> July 26, about 5 o'clock in the afternoon, while sitting in my dining-room at this place, Hastings, which is situated on the Parade, close to the sea shore, nearly fronting the south, my attention was excited by a great number of people running down to the sea side. On enquiring the reason, I was informed that the coast of France was plainly to be distinguished with the naked eye. I immediately went down to the shore, and was surprised to find, even without the assistance of a telescope, I could very plainly see the cliffs on the opposite coast; which at the nearest part, are between 40 and 50 miles distant, and not to be discerned, from that low situation, by the aid of the best glasses. They appeared to be only a few miles off, and seemed to extend for some leagues along the coast. I pursued my walk along the shore to the eastward, close to the water's edge, conversing with sailors and fishermen on the subject. At first they could not be persuaded of the reality of the appearance; but they soon became so thoroughly convinced, by the cliffs gradually appearing more elevated, and approaching nearer, as it were, that they pointed out, and named to me, the different places they had been accustomed to visit; such as, the Bay, the Old Head or Man, the Windmill, &c, at Boulogne; St Valery, and other places on the coast of Picardy; which they afterwards confirmed, when they viewed them through their telescopes. Their observations were, that the places appeared as near as if they were sailing, at a small distance, into the harbours.
>
> W. Latham

Similar views of France from the south-east coast of the British Isles occur regularly, though infrequently. For example, several people reported that the coast of France was clearly visible from Hastings on 5 August 1987.

It has been argued that looming led directly to the Norse settlement of Iceland because Norsemen were able to see some of its coast as a superior mirage from the Faeroe Islands. Later, Icelanders may have been similarly aided in their discovery of Greenland. Although most of the light from these distant shores is either scattered or absorbed by the atmosphere before it reaches the distant observer, the exceptionally good visibility brought about by the high-pressure conditions that favour the effect would just make them visible at distances of several hundred kilometres. Such extraordinary visual range is often noted by sailors. On 17 July 1939, a Captain John Bartlett clearly saw and identified the outlines of the Snaefells Jokull, a mountain 1430 m high situated on the western coast of Iceland, a distance of more than 500 km north-east of the position of his ship.

Chapter 4 | SUNSET AND SUNRISE

> It was one of the usual slow sunrises of this time of year, and
> the sky, pure violet in the zenith, was leaden to the northward,
> and murky to the east, where over the snowy down or ewe-
> lease on Weatherbury Upper Farm, and apparently resting upon
> the ridge, the only half of the Sun yet visible burnt rayless, like
> a red and flameless fire over a white hearthstone. The whole
> effect resembled sunset as childhood resembles age.
>
> Thomas Hardy, *Far from the Madding Crowd*, New Wessex
> Edition, Macmillan, 1977

4.1 Sunset

Are there significant differences between sunrise and sunset? It is a question
I often ask myself, and to which I have yet to find a definitive answer, if
indeed there is one. Other than that events in one occur in the reverse order
to events in the other, it seems to me that any physical differences are those
of detail only. By and large, it seems the conditions that give rise to a splendid sunset are exactly the same as those that produce an equally memorable
sunrise, although it has been said that these conditions occur more often at
the end of the day. Most of us have seen the Sun set many more times than
we have seen it rise, and so it's difficult to judge the truth of this. I invite
you to join me in watching sunsets and sunrises and put the claim to the
test. To avoid repetition, only sunsets are considered in what follows.

The colour and brightness of the Sun's disc are determined by the depth
of atmosphere through which its light must pass to reach our eyes. For much
of the day, especially during summer, when the Sun rides high in the sky, its
colour and brightness remain more or less constant because it is sufficiently
high in the sky that its rays encounter relatively few molecules on their way
to the ground. But in the late afternoon, as the Sun approaches the horizon,
its light reaches the ground through an ever increasing depth of air. The solar
disc grows dim, and its colour changes noticeably before our eyes. Its dulled
and reddened rays bathe our surroundings in a gentle rosy hue, and an amber
glow suffuses the western horizon. There is a feeling of transition, from
hectic day to tranquil night. This is one of the supreme moments of the day.
Yet it is soon over, a prelude to a lingering twilight glow that may last for an
hour or more after sunset. And whereas little effort is needed to notice the

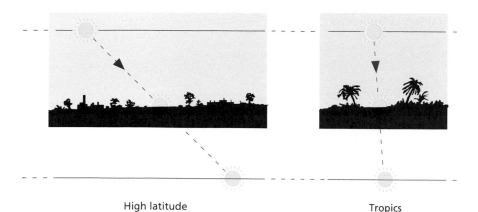

High latitude   Tropics

Figure 4.1 Duration of twilight. The Sun moves across the sky at a constant rate due to the Earth's spin. The duration of twilight depends on how long it takes the Sun to get sufficiently below the horizon so that its light no longer reaches the observer's sky, even indirectly. The diagrams show that this takes less time in the tropics, where the Sun's path is steeply inclined to the horizon, than it does at higher latitudes.

brief and showy blaze of a sunset, the subtle transformations of a twilight sky require an altogether more patient, practised eye.

The boundary between day and night is known as the terminator. Because the Earth rotates, every point on its surface passes through the terminator twice a day: once at sunrise, and again at sunset. On an airless world like the Moon, the terminator is well defined. Here on Earth sunlight is scattered by the atmosphere into the night sky long after the Sun has disappeared below the horizon, and the terminator is correspondingly poorly defined and diffuse. This scattered light is responsible for twilight.

If you have been to the tropics you may have been taken aback by how quickly twilight gives way to night. Twilight usually lasts longer at high latitudes than it does in the tropics because the Sun takes more time to drop a given number of degrees vertically below the horizon when the angle between the ecliptic and the horizon is small. In the tropics, the Sun's path is always steeply inclined to the horizon, and the onset of night tends to occur sooner and more suddenly than it does at higher latitudes.

Twilight is officially divided into three periods. The first of these is known as civil twilight, and it lasts from sunset until the brightest stars begin to be visible. This occurs when the Sun is at least 6° below the horizon. Civil twilight is followed by nautical twilight, which ends when the boundary between sea and sky is no longer discernible. The Sun is then at least 12° below the

(a) Sun at the horizon

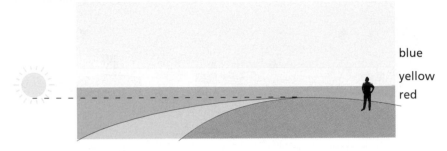

blue

yellow

red

(b) Sun below the horizon

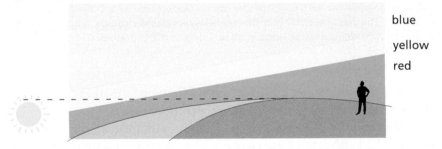

blue

yellow

red

Figure 4.2 Sky colours at sunset are due to selective scattering. This is greatest for sunlight closest to the ground, which is why the horizon looks red when the Sun is at or just below the horizon (diagram a). The sky just above the horizon will thus take on a pronounced reddish tinge. Above this, the sky fades to yellow and higher still to blue.

As the Sun drops below the horizon the reddish band grows wider because the only light illuminating the horizon is reddish (diagram b).

horizon. Finally there is astronomical twilight, which lasts from the end of nautical twilight until the faintest stars that can be seen by the naked eye are visible. This occurs when the Sun has sunk 18° below the horizon.

The end of astronomical twilight marks the beginning of night, when no part of the atmosphere visible to you is illuminated by the Sun, even indirectly. At high latitudes (i.e. more than 48.5° N or S), however, in midsummer the night is never completely dark because astronomical twilight lasts from sunset to sunrise. The faintest stars are not visible in the twilight glow, which makes stargazing in summer at high latitudes far less rewarding than it is in winter, even when you are far from the urban glow.

Figure 4.3 Sunset. A multiple exposure of sunset. When the Sun is close to the horizon its disc is reddened and distorted into an oval. (*Photo* Pekka Parviainen)

When you have watched the Sun set a few times you begin to realise that during sunset, and the twilight that follows, the appearance of the sky and landscape changes in a more-or-less predictable manner. But the sheer number and subtlety of these changes means that you have to watch many sunsets before you are familiar with all the nuances. And since twilight lasts an hour or more after the Sun has disappeared below the horizon, you have to be in a particularly patient frame of mind if you are to see the spectacle through till the Earth's rotation has swept you entirely away from the last vestiges of sunlight.

Ideally, the sky should be completely free of clouds and haze, both above and below the horizon, and you should have an unobstructed view of both western and eastern horizons. This way you get to see the spectacle in its purest form. Clouds do make for a spectacular sunset, but get in the way of the twilight that follows.

A great place from which to see sunset is a plane that is flying eastward during sunset. In the first place, viewing conditions are ideal: you are usually far above any clouds, and you have an unobstructed view of the horizon. As a bonus, sunset and twilight are speeded up because you are travelling away from the Sun, so that the whole show can be over within half and hour. Be

Figure 4.4 Sunset clouds. In the best sunsets the sky is filled with clouds that reflect the Sun's reddened light. The photograph at the top shows sunset-illuminated clouds beyond Zugspitze (the highest mountain in Germany). The one at the bottom was photographed from South London. (*Top photo* Claudia Hinz; *bottom photo* John Naylor)

sure to pick a window seat on the side of the plane that will give you a view of the Sun. In the northern hemisphere this will on the right-hand side.

Now to the mechanics. Imagine that you are looking westward at the end of the day. The air is clear, and there are few clouds. Twilight colours first become noticeable about half an hour before the Sun reaches the horizon. Although most of the sky is still blue, and the Sun is surrounded by the characteristic colourless glow of its aureole, faint traces of sunset colours begin to appear along the horizon: pale yellow in the west and a rosy hue in the east. Avoid looking at the Sun: apart from possible harm to your eyesight, the Sun is so much brighter than the rest of the sky that it blunts the sight, and makes it difficult to notice the subtle sunset colours.

As the Sun sets, and sometimes just before it sets, a blue-grey band appears along the eastern horizon. This band is commonly known as the 'Earth's shadow', although, for reasons given below, it should really be called the 'dark segment'. The faint rosy fringe above the dark segment is the anti-twilight arch, or counterglow, sometimes called the Belt of Venus. As the Sun drops further below the horizon, the dark segment grows broader and less distinct, vanishing when its upper edge is between 5° and 10° above the horizon. The best views I have had of the dark segment and the counterglow have always been from a plane.

After the Sun has set, its light continues to illuminate the sky above you. Twilight has begun. Along the western horizon, this sunlight is visible as a shallow, horizon-hugging arch that extends as much as 90° either side of the point at which the Sun has set. This twilight arch, as it is known, grows fainter away from the Sun, and changes colour as the Sun sinks further below the horizon. Directly above the horizon, the twilight arch is orange, fading to yellow higher up. Occasionally, the upper boundary of the twilight arch takes on a faint sulphurous-green hue. Try watching the evening sky while lying on your side: curiously this emphasises the coloured bands.

About 15 minutes after sunset, you can sometimes catch the first glimpse of the 'purple light'. This is a faint pinkish or lilac patch of light visible above the twilight arch. The 'purple light' reaches its maximum brightness, which is seldom very bright compared with other sunset colours, about half an hour after sunset, i.e. at the end of civil twilight. After this, colour drains from the sky, and all that remains is a very pale, colourless twilight glow which gradually shrinks into the western horizon at a rate that depends on both latitude and the season. In the east, the dark segment ceases to be discernible some 20 to 30 minutes after sunset, leaving the eastern sky uniformly dark grey. The very last traces of twilight disappear about an hour and a half after sunset. Night has finally fallen.

Throughout the whole twilight period the brightest parts of the sky are just above the eastern and western horizons. The zenith sky, which remains blue, becomes increasingly dark.

From the moment that the Sun sets, the entire landscape is illuminated only by scattered light. Scattered sunlight is, of course, noticeably blue. The absence of direct – i.e. white – sunlight after sunset means that blue hues become much more intense during twilight. For example, blue flowers look much bluer than they do in direct sunlight. At the same time, shadows cast by artificial lights such as sodium street lamps or household filament lamps also take on a distinctly blue hue. For an explanation of coloured shadows see section 2.5.

4.2 Twilight

The atmosphere robs Peter to pay Paul. But in this case everyone is a winner. A glorious amber sunset is achieved by stripping sunlight of much of its blueness, and redirecting this blue light into someone else's late afternoon sky.

Most of this happens close to the ground. The amount of sunlight scattered by the atmosphere decreases with increasing altitude because the density of the atmosphere drops rapidly with increasing height. At heights of more than 50 kilometres above the Earth there are too few molecules to scatter significant amounts of sunlight. Twilight is therefore brought about in, and confined to, the lowest levels of the atmosphere.

Sunlight that skims the Earth's surface at the terminator is largely shorn of its shortest wavelengths through selective scattering by the high concentration of molecules of gas and other tiny particles that make up the air close to the ground. Sunlight thus becomes progressively reddened as it penetrates the atmosphere on its way to your eye. You see this light as a narrow orange strip along the western horizon, which corresponds to the depth of atmosphere in which selective scattering is greatest, and as a rosy fringe above the eastern horizon. The pale yellow band above the orange one is due to light that passes through the atmosphere a few kilometres above the Earth's surface. Here it encounters fewer molecules and so retains a larger proportion of its shortest wavelengths. Sunlight that passes through the atmosphere at even greater altitudes undergoes little scattering and thus remains white. This unscattered light is responsible for the colourless twilight glow high above the western horizon. The blueness of the zenith after sunset is due in part to light scattered from sunlight that passes through the atmos-

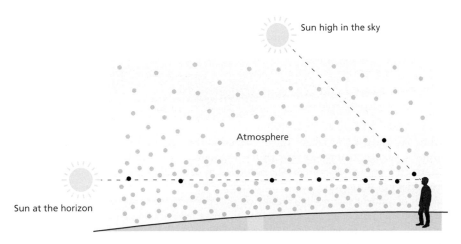

Figure 4.5 Causes of twilight colours. Rayleigh scattering is greatest close to the ground where there are most molecules. Furthermore, when the Sun is at the horizon its light must pass through a far greater amount of atmosphere than when the Sun is high in the sky. Taken together, these reduce the proportion of short wavelengths (responsible for the blues and violets in sunlight) that reach the sky around you at sunset and sunrise. When the Sun is high in the sky selective scattering has less effect because sunlight passes through less atmosphere to reach you and so encounters fewer molecules.

phere high above you. Absorption of longer wavelengths within the ozone layer, which is at an altitude of some 20 kilometres above the ground, also contributes to the blueness of the zenith sky after sunset.

What happens to the shorter wavelengths that fail to reach your western horizon? Well, they are scattered to give people west of you their blue afternoon sky. At the same time, your sunset is seen as a sunrise by someone on the other side of the Earth: for example, sunset in Europe almost coincides with sunrise in New Zealand, and vice versa.

Another way to understand the changes that are seen in the sky after sunset is to picture twilight as a broad beam of sunlight made up of horizontal layers of different colours (see figure 4.2). The layer furthest from the Earth's surface is the least affected by scattering and so is composed of white light. The layer nearest the Earth is most affected by scattering and so is reddish. Sandwiched between them is a layer of yellowish light. At the moment of sunset this beam is parallel to the horizon, and just glances the Earth's surface at the terminator. After sunset, the Sun's light can shine directly only into the sky above you, and so the lowest layer of the beam illuminates the atmosphere above the eastern horizon. As the Sun drops further below the horizon, its rays enter your bit of the sky at an ever steeper

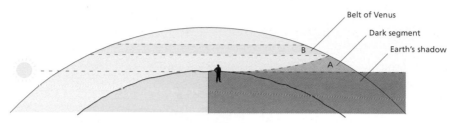

Figure 4.6 The Earth's shadow. The brightness of the sky is due to scattered sunlight. However, sunlight that passes through the atmosphere is gradually reduced in intensity due to scattering and absorption. The reduction in intensity is greatest in sunlight that passes through the lowest levels of the atmosphere. When the Sun is on the horizon the amount of sunlight that reaches the sky on the opposite side of the atmosphere (A) has been reduced to the point that the horizon in this direction looks grey. This grey band is known as the dark segment. The pinkish band (B) above the dark segment is known as the 'Belt of Venus' and is due to sunlight that has lost much of its bluish light.

angle, so that, 20 to 30 minutes after sunset, sunlight no longer reaches the eastern horizon directly. Eventually only the sky high above the western horizon is directly illuminated by it. Throughout this period, the horizon remains orange or red because, although the Sun can't illuminate the horizon directly, its red light is scattered in the forward direction by the atmosphere, and so some light reaches the western horizon.

The dark band along the eastern horizon is a consequence of the large amount of sunlight that is scattered as it passes through the atmosphere close to the Earth's surface. This greatly reduces the amount of sunlight that reaches the atmosphere above the eastern horizon, which therefore appears both dark and colourless. The reduction in brightness would not be as noticeable were it not for sunlight scattered back in the direction of the observer from the atmosphere above the dark band. This light, the shorter wavelengths of which are scattered as it passes several kilometres above your head, forms a rosy border to the dark band. However, as the Sun drops further below the horizon, its light can illuminate only the upper part of the atmosphere in the eastern sky. Eventually it shines on air that is too tenuous to scatter light, and the rosy fringe disappears, rendering the dark band indistinguishable from the rest of the now-darkened eastern sky.

Strictly speaking, since it is due to scattering, the dark band seen in the east is not really the Earth's shadow, although it is sometimes called this. Furthermore, it is actually slightly broader than the Earth's shadow would be, though its maximum breadth seldom exceeds 5° above the horizon for the reasons given above. As the Sun drops ever further below the western horizon, the dark segment sweeps across the sky from east to west, and so passes over your head. You won't notice its overhead passage, since not enough light is scattered from the zenith to define its edge.

The rosy light of the anti-twilight arch (the Belt of Venus) is also responsible for the 'alpenglow', the rosy glow that is sometimes seen on the snow-covered peaks of tall mountains in the Alps shortly after sunset. The alpenglow disappears when the peaks are no longer directly illuminated by the Sun. Occasionally, the purple light (see section 4.1) gives rise to a second alpenglow.

4.3 Clouds at sunset

Clouds are not themselves the cause of sunset colours. They can, however, make all the difference to a sunset by reflecting the rosy hues of the Sun just after it has set. Arguably, the most eye-catching sunsets are those in which the sky is covered with ribbons of altocumulus cloud. Sunlight reflected by these clouds not only creates a glorious spectacle in the sky, it also bathes your surroundings in a diffuse amber glow. Without a layer of clouds to scatter the Sun's reddened light, this amber glow is absent from a sunset.

If you look around a sky that is dotted with clouds during sunset you will notice that, at any particular moment, only some of them are tinged with sunset colours. The rest are either white, or various shades of grey. Furthermore, there is a pattern to the way in which clouds become tinged with colour during sunset. Just before sunset the reddened light that we see on the horizon illuminates the atmosphere closest to the Earth's surface.

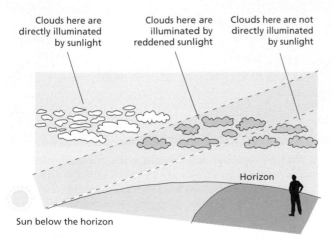

Clouds here are directly illuminated by sunlight

Clouds here are illuminated by reddened sunlight

Clouds here are not directly illuminated by sunlight

Horizon

Sun below the horizon

Figure 4.7 The colour of clouds at sunset. This depends on where they are in the observer's sky. Those furthest from the ground are the last to be directly illuminated by the Sun and will appear white. At the same time clouds closer to the observer are illuminated by the Sun's reddened light and appear pinkish. Clouds on the opposite side of the sky from the Sun are illuminated by only by skylight and look grey.

Figure 4.8 Earth's shadow. The top photograph was taken from the ground, the one at the bottom was taken from a mountain at an altitude of 1835 m. In both, the dark strip along the horizon is the Earth's shadow and the pink band above it is the Belt of Venus. In the bottom photograph you can see a rising Moon. (*Top photo* John Naylor; *bottom photo* Claudia Hinz)

Clouds close to the ground take on a rosy hue, while those at a greater height remain uncoloured because they are illuminated by light that is still white. As the Sun sinks below the horizon, its reddened light no longer reaches clouds on the eastern horizon, and so these turn various shades of grey, while clouds directly above you are either pink or white depending on how high they are. Clouds in the east no longer illuminated by direct sunlight can take on a blue-grey hue if they are illuminated by scattered skylight, which is, of course, predominantly blue. Just after sunset, clouds directly above you that are closest to the ground become grey, while higher ones turn pink. Later, these also turn grey. The last clouds to change colour are those on the western horizon. Tall clouds can have a rosy top and a grey base, or a white top and a rosy base.

Clouds sometimes clear from the sky after sunset because the Sun's heat that keeps them fuelled is no longer available. There are, however, occasions when this disappearance is just an illusion. Wisps of cirrus often appear to vanish from view after sunset because they are no longer illuminated directly by the Sun, and so are hardly any brighter than the surrounding sky. You can make them reappear with a polarising filter by rotating it to reduce the brightness of the background sky.

4.4 The purple light

On some evenings – sometimes for several successive evenings – you may just be able to notice a faint lilac tinge in the twilight glow above the western horizon. This is the purple light. It usually makes its appearance just above the twilight arch some 15 or 20 minutes after sunset, and spreads throughout the twilight glow over the next quarter of an hour, reaching its maximum brightness towards the end of this period – i.e. about half an hour after sunset. It then fades away, and the last traces may be visible half an hour after you first noticed it, though it may sometimes last longer.

The purple light is usually so faint that it is difficult to see if you look at it directly. The eye can't discriminate colours very well when the level of illumination is low. Hence, you should rest your eyes frequently when looking at or for the purple light, either by closing them from time to time for half a minute or so, or by turning your back on the western sky for a minute or two. When looking for the purple light, glance out of the corner of your eye.

The brightness of the purple light varies from one evening to another. Often it is absent. The purple light is brightest after volcanic eruptions,

Figure 4.9 Purple light. This is a photograph of a sunset taken in London, England in 1992. The colour of the twilight arch is due to particles emitted by Mount Pinatubo which had erupted several months earlier the previous year. (*Photo* John Naylor)

though not every eruption leads to an increased purple light. Figure 4.9 shows the purple light that followed the eruption of Mt Pinatubo in 1991. It was easily discernible to the naked eye for many weeks in November and December of that year in northern Europe.

The purple light is thought to be due to a permanent layer of haze particles, mainly drops of sulphuric acid, estimated to lie between 15 and 20 km above the Earth's surface. The concentration of particles in this layer varies from season to season, and is greatest following some, though not all, volcanic eruptions. These particles scatter reddened sunlight, forming a rosy veil between you and the scattered blue of the twilight in atmosphere beyond the dust layer. The eye is unable to separate out these colours, and sees the combination of blue and red as a pinkish purple.

Since the red component of the purple light is contributed by sunlight that has skimmed the Earth's surface, large clouds and tall mountains beyond the western horizon can get in its way. This produces gaps in the veil of rosy light, through which you see the blue skylight beyond as faint blue-grey streaks that radiate from the horizon where the Sun has set. These streaks are known as crepuscular rays. Because they are blue, crepuscular rays are difficult to see when there is little or no purple light. The blueness of crepuscular rays is made even more marked by simultaneous colour contrast because blue and red are complementary colours.

4.5 Crepuscular rays

The word crepuscular comes from the Latin for twilight. Crepuscular rays are not uncommon, though they frequently go unseen because the difference in brightness between them and the surrounding sky is often so small that if you are inexperienced or impatient you won't notice anything unusual. Nevertheless, if you watch sunsets regularly until the end of civil twilight, expect to see one or more faint crepuscular rays on several occasions.

Crepuscular rays are no different from the shadows cast by clouds when the Sun is above the horizon. Like them, they appear to fan out from their source because of perspective. What makes them visible is sunlight scattered by the air in the gaps between the shadows. Scattering itself depends on a number of factors: the size and concentration of scattering particles and the direction from which they are illuminated. Only shadows cast into the lowest levels of the atmosphere are visible because at increasing altitude the density of scattering particles is too small to produce the contrast necessary to make the shadow visible.

Crepuscular rays stand out particularly clearly in a sky in which you can also see the purple light. Occasionally, a combination of a large number of crepuscular rays and a very strong purple light produces the most spectacular twilight display imaginable.

Several conditions must be satisfied for crepuscular rays to occur. The most important of these is the size and position of the object casting the shadow. It has to be large enough to cast an umbral shadow that can reach the sky above your horizon and it has to lie at or very near the terminator

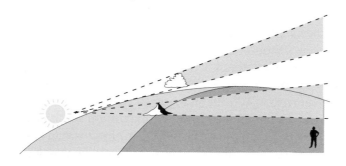

Figure 4.10 Crepuscular rays. These are vast shadows cast into the sky after sunset, or before sunrise by huge clouds or mountains several kilometres beyond the horizon. Perspective makes them appear to spread out from the objects casting the shadows.

Figure 4.11 Crepuscular rays, photographed in Nepal and caused by mountains. (*Photo* Claudia Hinz)

so that it gets in the way of the sunlight that grazes the Earth's surface. At the same time the sky must be sufficiently dark for the shadow not to be overwhelmed by the sky's brightness. This means that an object has to be several hundred kilometres beyond your horizon if it is to give rise to crepuscular rays that you will see. This, in turn, imposes a limit on the minimum size of the object because if it is to cast an umbral shadow it has to have an angular diameter that is greater than that of the Sun measured from the point from which you see the shadow.

Since the Sun's angular diameter is half a degree, a cloud or mountain must be at least 1 kilometre across if it is just to cover the Sun for someone 120 km away. However, when the terminator is 120 km from you, the Sun is only about 1° below the horizon, and the sky will usually be too bright for crepuscular rays to be seen. Crepuscular rays usually become visible when the Sun is between 3° and 4° below the horizon, which places the terminator 300–450 km beyond the horizon. The rays usually cease to be visible when the terminator is more than 700 km away, when the Sun is 6° below the horizon, and civil twilight has ended. Since the maximum reach of an umbral shadow is 120 times the diameter of the object, an object would have to be at least 6 km wide to cast an umbral shadow that would be visible at this distance. Don't expect to see crepuscular rays in a sky in which you can just begin to make out stars.

On rare occasions you may see crepuscular rays converging on the anti-solar point (see section 2.1) at the eastern horizon. These crepuscular rays are also known as anti-crepuscular rays. Sometimes they are due to per-spective convergence of the crepuscular rays originating below the opposite horizon. At other times they are caused by clouds that are above the horizon. There is more on anti-crepuscular rays in section 5.12. Exceptionally, you may see crepuscular rays arching overhead from one horizon to the other.

4.6 Mountain shadows

When you see a crepuscular ray you are at the receiving end of the shadow of a huge, distant object. Conversely, if you are standing on top of a moun-tain at sunset or sunrise, you may see its shadow cast towards the antisolar point. The remarkable thing about mountain shadows is that, seen from the top of the mountain, they all appear to have more or less the same shape. Regardless of the mountain's profile, as far as can be judged by the naked eye, its shadow is invariably triangular, and its apex lies at the antisolar point. The appearance is due to perspective convergence. The shadow cast by the top of the mountain when the Sun is on the horizon falls so far away that it appears almost as a point to the eye.

4.7 Abnormal twilights

To make the Sun look distinctly red a much greater concentration of scat-tering particles is necessary than is to be found in a clean, clear sky. These extra particles are supplied by microscopic fragments of various materials that are taken up into the atmosphere, and by solution drops formed when humidity is high enough to allow water vapour to condense on particles with a strong affinity for water. City dwellers probably see highly coloured sunsets more often than those who live in the country because city air is more likely to be heavily laden with suitable scattering particles. These par-ticles are found in exhaust emission from cars, factories and fireplaces. The fact that man-made pollution is occasionally responsible for one of nature's more beautiful sights rather robs city sunsets of their magic.

The most spectacular sunsets tend to occur following cataclysms such as massive volcanic eruptions, or a large meteoroid striking the Earth. These events can throw a vast amount of material in particle form into the atmosphere to heights of 20km or more, where it can remain for months or even years. The eruption of Krakatoa in 1883 was followed for several months by spectacular sunsets that were seen all around the world. There were several volcanic eruptions during the twentieth century that were violent enough to affect sunsets in this way. Highly coloured sunsets were seen following the eruption of Mount Saint Helens in the USA in 1980, and El Chinchon in Mexico in 1982. Most recently, the eruption in 1991 of Mt Pinatubo in the Philippines led to many wonderful sunsets in Europe later that year. Spectacular sunsets were seen in northern Europe for several weeks after the so-called Tunguska event, which is now believed to have been a very large rocky meteoroid that exploded over Siberia in 1908.

Twilight in volcanic sunsets is brighter and much more highly coloured than at other times. One of the hallmarks of these sunsets is a particularly bright purple light. The most recent example of bright purple light sunsets occurred a few months after the eruption of Mt Pinatubo in mid-1991. These sunsets were seen in the USA and northern Europe.

4.8 The Sun at the horizon

None of the effects mentioned so far rely on the refraction that occurs as the Sun's light passes through the atmosphere. The degree of refraction increases as this light approaches the Earth's surface because of the increasing density of the atmosphere. The resulting change in direction of the rays is minute, though measurable, and is most marked in the rays that strike the atmosphere obliquely, i.e. in those from the rising and setting Sun. Consequently, the most noticeable effects of atmospheric refraction on the shape and colour of the Sun occur at sunset or sunrise. The appearance of the Moon's disc is also noticeably altered by atmospheric refraction as it approaches the horizon.

A change in the direction of the rays by which an object is seen may bring about an apparent change in both its shape and position. This is why mirages are seen. Where the Sun is concerned, its disc is noticeably flattened when it is on the point of rising or setting because light from its upper edge is refracted slightly less than light from its lower one. A disc that is flattened is said to be oblate. The Sun's oblateness is even more pronounced if you cock your head, and look at it sideways. The different vantage point accentuates the effect: try this with figure 4.3.

Refraction can also significantly alter the apparent position of the Sun.

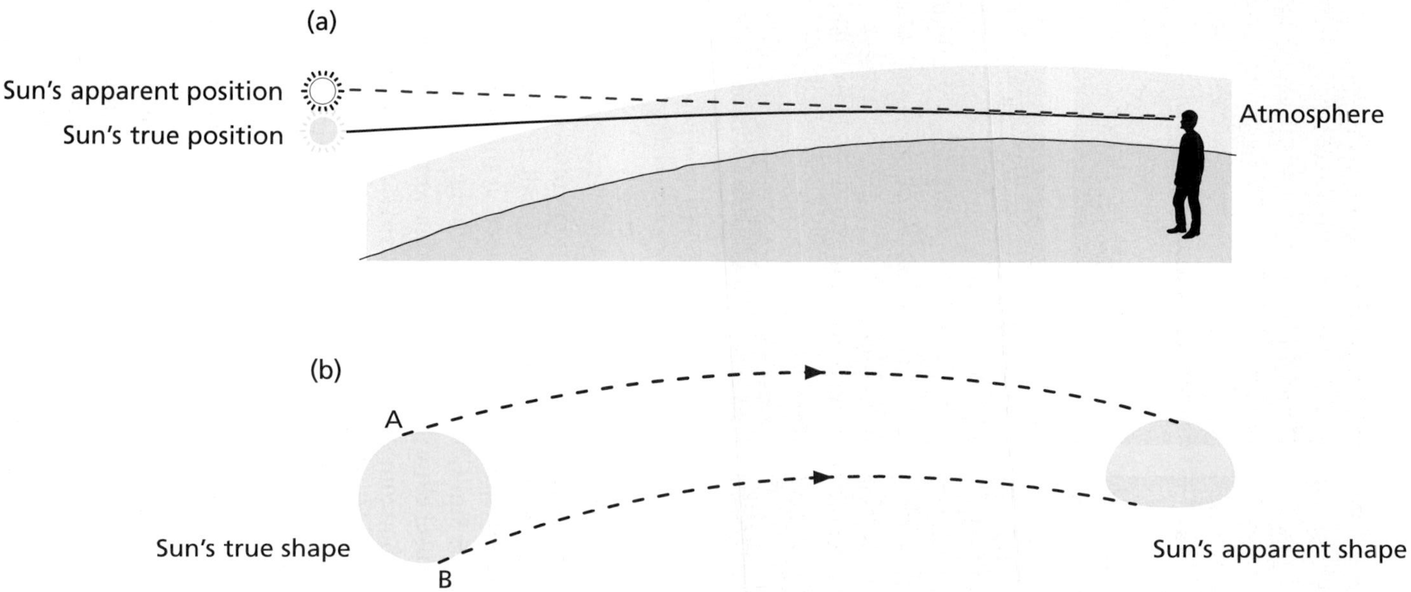

(a)

Atmosphere

Sun's apparent position

Sun's true position

(b)

A

B

Sun's true shape

Sun's apparent shape

Figure 4.12 Atmospheric refraction at sunset. Sunlight is refracted as it makes its way through the atmosphere and follows a path that is concave to the Earth's surface. Consequently, when you see the Sun on the horizon it has already set (a). At the same time when the Sun is at the horizon, rays from the top of its disc (A) are not as strongly refracted as they pass through the atmosphere than those from the bottom of its disc (B). This makes the Sun appear as if it has been squashed (b).

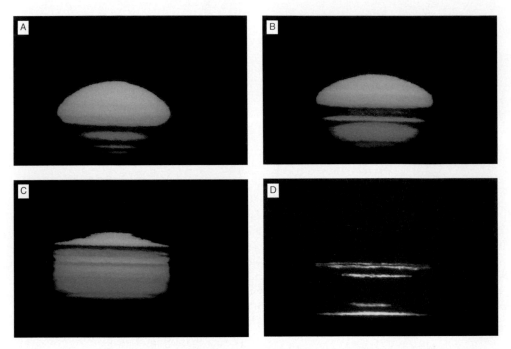

Figure 4.13 Setting Sun sequence. These photographs (A, B, C, D) show how dramatically the Sun can change shape as it sets when there is a strong temperature inversion within the atmosphere close to the ground. (*Photo* Pekka Parviainen)

When the Sun is at the horizon, atmospheric refraction brings about a vertical shift in its position that amounts to approximately half a degree. This is the same as the Sun's apparent angular diameter. Hence, at the very moment when the lower edge of the Sun's disc appears to touch the horizon, the Sun itself is actually entirely just below the horizon. In other words, when you see the setting Sun reach the horizon, the Sun itself has physically already set.

The Sun's shape can be further distorted if it is seen across a warm surface, or through a strong temperature inversion a few metres above the ground. As we have seen, these are the conditions that favour the mirage. The apparent change in the Sun's shape depends on the temperature profile of the atmosphere. On occasion, when the brightness of the Sun's disc is sufficiently diminished by haze so that you can look at it without being dazzled, some of these distortions are noticeable to the naked eye.

It's worth repeating at this point that you should always avoid looking directly at the Sun when it's above the horizon. However, it's usually safe to look at the Sun briefly when it is on the point of setting because scattering diminishes its brightness to a safe level. If you can look at its disc comfort-

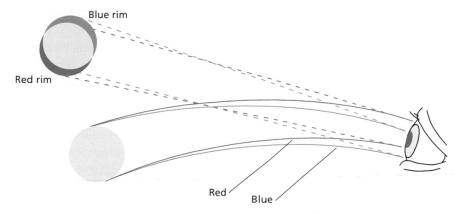

Figure 4.14 As sunlight passes through the atmosphere it is very slightly refracted. The amount of refraction depends on the frequency of light, and so the Sun's white light is very slightly separated into its component colours. Bluish light undergoes slightly more refraction than red light, as shown in the diagram. The effect is greatest when the Sun is at the horizon. However, instead of a seeing a bluish tinge along the upper edge of the Sun you are much more likely to see a greenish one because blue light is scattered out of the Sun's beams to a greater extent than green. The bottom of the Sun will be reddish.

ably, then it's not bright enough to harm your eyesight. However, to see the finer details, you need to use binoculars or a low-power telescope, both of which should be equipped with a suitable neutral density filter. A neutral density filter reduces brightness without altering colour.

The commonest distortion occurs when the Sun rises or sets over a large body of water that is warmer that the air above it. In these circumstances refraction may bring about an inferior mirage of the Sun. As the Sun approaches the horizon you will see a segment of its inverted image rising from the horizon to merge with it. From the moment when the mirage becomes visible, it takes only a few seconds for the Sun to merge with its mirage, forming a shape like the Greek letter omega (Ω) (figure 3.4).

An inversion a few metres above the Earth's surface brings about another type of distortion in which the Sun takes on shapes such as those shown in the photographs (figure 4.13).

Refraction is a colour-dependent phenomenon: the degree of refraction is inversely related to the wavelength of light. Thus red light is refracted very slightly less than blue light as it passes from one medium to another. Although this difference is extremely small where the atmosphere is concerned, it is nevertheless sufficient to cause colour-separation at the edges of the Sun's disc when the Sun is at the horizon. Under some circumstances this colour separation is noticeable to the naked eye as a brief flash of green light (see section 4.9).

The separation of a beam of light into its constituent colours by refraction is known as *dispersion*. The effect of dispersion by the atmosphere on the Sun's disc is not directly visible to the naked eye because the amount of dispersion is less than the eye's ability to resolve it, i.e. less than one minute of arc, $\frac{1}{60}$ degree. In any case it is only at the top and bottom of the Sun's disc that dispersion is noticeable. Over the rest of the disc, dispersed colours overlap, and most of the disc would appear white were it not for selective scattering by the atmosphere. When the Sun is on the horizon selective scattering removes much of the shorter wavelengths. Depending on the amount of scattering, which increases if there is a haze, the disc appears yellow or orange.

When the Sun is on the point of setting, you can sometimes see with binoculars a distinct red fringe along the lower edge of the Sun's disc. This fringe is due to dispersion, and not to scattering. It is much more difficult to see the dispersed image along the upper edge of the Sun's disc. On the basis of dispersion alone this should be blue. This colour is usually not seen because a large proportion of blue light is scattered out of sunlight as it passes through the lower atmosphere. However, you can sometimes see a blue rim if you are a few kilometres above sea level, say in an aeroplane. The Sun is then seen through a much more tenuous atmosphere than at sea level, and so a far smaller quantity of the shorter wavelengths are scattered out of its light. At sea level, even when the atmosphere is very clean, the Sun's upper rim is likely to be green. This green rim is much too narrow to be seen by the naked eye, though it can be seen with binoculars.

Scattering is also responsible for the noticeable diminution in the brightness of the Sun's disc when it is within a few degrees of the horizon. Scattering reduces the apparent brightness of any celestial object because its light must pass through the atmosphere. Astronomers call this effect 'atmospheric extinction' (see section 12.4). When the Sun is at the horizon, even in clear air, its brightness may be as much as 9 magnitudes less due to extinction than when it is within 5° of the horizon. A difference of 9 magnitudes is a change of approximately 3800 times in brightness. If there is a haze, the reduction in brightness will be even larger.

4.9 Green flashes

Although the blue and green rims of a setting Sun are much too narrow to be seen by the naked eye, green flashes have been seen countless times

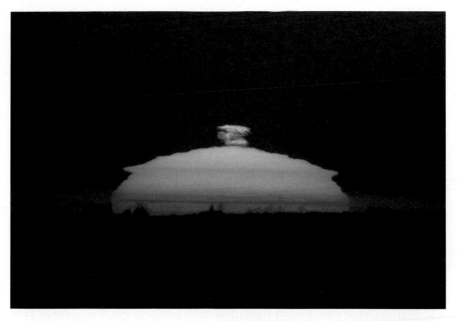

Figure 4.15 Green segment. The Sun's green topknot in this photograph is the topmost segment of the Sun's disc distorted by atmospheric refraction. (*Photo* Pekka Parviainen)

without the aid of a telescope or binoculars. Even blue flashes have been seen, though far less often than green ones, and only from altitudes of several kilometres.

The key to seeing a flash is that the green or blue rim must either be seen in isolation from the rest of the Sun's disc, or be magnified, or both. Your chances of seeing a green flash are improved if you use binoculars, though, arguably, it should be visible to the naked eye if it's to be worthy of the name. A flash suggests something bright enough to draw attention to itself without the need for artificial aids.

Isolation of the upper edge of the Sun is easily arranged: it happens at every sunrise and sunset during the few seconds that it takes the edge of the Sun's disc to rise above or set below the horizon. But if the air isn't clear, scattering of short wavelengths by the atmosphere will reduce the brightness of the green rim below that necessary to see its colour. Magnification of the green segment can be brought about either by inferior mirage conditions, or by a temperature inversion some distance above the Earth's surface.

Inferior mirage conditions are the most common cause of the green flash. Look for a tell-tale distortion in the Sun's disc when it is almost touching the horizon: in inferior mirage conditions, as we have seen, the Sun merges with its mirage to resemble the Greek letter Ω. When the upper edge of the Sun, its green rim, is close to the vanishing line of the mirage, it merges with a mirage of the green rim, so that all that you see is a green spot. With luck this may be large enough and bright enough to be seen with the naked eye for the second or so before the Sun either slips below the horizon or rises above it.

A temperature inversion can also magnify the green rim. As in the case of an inferior mirage, the green flash will only be visible to the naked eye if it is seen in isolation. On other occasions a green flash has been seen when the Sun is close to the horizon and its disc is hidden behind a cloud, hill or mountain.

A green flash is not, in fact, always green. It can appear green to your eye because you have been staring at a reddish Sun. Doing so makes your eye much less sensitive to longer wavelengths, and light that would otherwise look yellow is seen as green. This shift in the eye's perception of colour explains why in many photographs a green flash appears yellow, and not green as it did to the eye of the photographer.

Even if conditions are just right, you won't see a green flash unless your eye is above the horizon. If you are standing on a beach, or you are in a boat, the horizon will be below eye level. This is why most green flashes are seen over water. Alternatively, look from the top of a cliff, a very tall building, or from an airborne plane.

4.10 The Purkinje effect

A curious effect can be noticed at twilight, or whenever the intensity of illumination is greatly reduced. The effect is named after the man who first brought it to public attention. His name was Evangelista Purkinje, and he was the Professor of Physiology at the University of Prague until his death. He discovered many interesting eye-related optical effects, one of which has come to be known as the Purkinje effect.

He noticed that, when the level of illumination is reduced, red objects appear to be much darker than those that are green or blue, even when, under normal illumination, the reverse is the case. You may have noticed this occasionally when out of doors some while after the Sun has set: a green lawn appears to be much brighter than the earth in the flower beds, or the wooden fences that surround them, whereas earlier, when there was more light, they all appeared equally bright.

The effect is a purely physiological one; it is due to the way in which our eyes work. The retina of each eye contains two distinct types of light-sensitive cells known as rods and cones. Rods are sensitive only to brightness and cannot distinguish colour, whereas cones are sensitive to colour and brightness. However, rods are most sensitive to light that lies in the blue-green part of the spectrum and cones are most sensitive to light that is apple green. As the intensity of light diminishes, the cones become insensitive and we cease to distinguish colours. At this point the rods take over vision. When this occurs, red and yellow objects no longer look as bright to us as those that are green or blue even though both are equally brightly illuminated. The Purkinje effect involves a change in intensity and not in colour.

The Purkinje effect is helpful if you need to read maps or use instruments at night while at the same time preserving your dark adaptation: you can use a red light as a source of illumination. The red light stimulates the cones (thus enabling you to see details clearly), but not the rods, which therefore retain their dark adaptation. Incidentally, cones are themselves not particularly sensitive to red light, so anything illuminated by red light does not appear very bright.

One of the curious things about this shift in sensitivity towards the shorter wavelengths is that the only major source of natural light at night, the Moon, is richer in long wavelengths than it is in short ones. Why, then, are our eyes more sensitive to shorter wavelengths in conditions of low illumination? What evolutionary pressures brought this about? I don't know the answer.

Chapter 5 | RAINBOWS

Shortly we came in sight of that spot whose history is so familiar to every school-boy in the wide world – Kealakekua Bay – the place where Captain Cook, the great circumnavigator, was killed by the natives, nearly a hundred years ago. The setting sun was flaming upon it, a Summer shower was falling, and it was spanned by two magnificent rainbows. Two men who were in advance of us rode through one of these and for a moment their garments shone with a more than regal splendor. Why did not Captain Cook have taste enough to call his great discovery the Rainbow Islands? These charming spectacles are present to you at every turn; they are common in all the islands; they are visible every day, and frequently at night also – not the silvery bow we see once in an age in the States, by moonlight, but barred with all bright and beautiful colors, like the children of the sun and rain. I saw one of them a few nights ago.

Mark Twain, *Roughing It*

5.1 Unweaving the rainbow

How many rainbows have you seen recently, say in the last six months? Not many, I'll wager. The fact is that, unless you live somewhere where the conditions that favour their formation occur on an almost daily basis, such as an oceanic island like Hawaii, you probably don't often get the chance to see a rainbow. In Hawaii, if Mark Twain is to be believed, they are so common that you can't miss them. But, wherever you live, rainbows almost certainly occur more frequently than you think. It's mostly a matter of looking for them when conditions are right.

Almost everyone knows the old rule of thumb that Sun plus rain equals a rainbow. But, as you may have discovered, this rubric isn't foolproof. What's missing are important details like the Sun must be close to the horizon and the rain must be in the opposite quarter of the sky. In fact, if you are a novice skywatcher, by the time you've taken on board all the extra conditions necessary to guarantee a rainbow, your initial enthusiasm may be quite blunted! But don't give up. When you've gone out of your way to see a few rainbows, and learned more about the circumstances in which they are likely to occur, you'll find that it becomes second nature to look at the sky whenever conditions seem right. Don't expect always to see a rainbow. That's part of their charm: these wonderful, candy-striped arcs are never completely predictable. You'll sometimes find that, even when it seems that all the necessary conditions have been met, no rainbow appears.

I'm taking for granted that you know what a rainbow is. But I'll hazard a guess and say that what you know about rainbows owes more to hearsay than to first-hand observation. As you leaf though this book you'll realise that rainbows are not the only coloured arcs that can be seen in the sky. Ice halos (section 7.1) and coronae (section 6.1) around the Sun are far more common than rainbows, and are often mistaken for them by people who are unfamiliar with the sky. Why don't these people know as much about halos and coronae as they seem to know about rainbows? There are several reasons.

People have always been fascinated by rainbows. We can deduce as much from mythology and folklore because stories about rainbows are strikingly similar the world over, which suggests that they hark back to an era, tens of thousands of years ago, before our ancestors began to wander across the globe. What was it about the rainbow that caught their imagination? Its colours? Its symmetry? Its transience? Perhaps all of these. But I think that what really struck them was something that always strikes me when I see a rainbow: the way it seems to link earth and sky. There is something decidedly otherworldly about it.

In a lot of folklore the rainbow is a bridge between this world and the next. The souls of the dead were said to cross over it into the afterlife. And shamans, people with the power to mediate between mortals and gods, claimed that they used it to communicate with the gods. One way or another, when people saw a rainbow they believed it to be a sign of divine activity, a message from the gods. Sometimes it promised reconciliation, like the rainbow that followed the Biblical flood. But, to judge from the many superstitions associated with rainbows, people were more concerned with its association with death and the afterlife. Many superstitions about rainbows, such as not pointing at it, or avoiding land where it touched the ground, were ways of avoiding its supposed malign associations with the next world.

Although some of this archaic symbolism lives on in painting and in poetry, the age of myth is over. To comprehend myth in the right way requires a pre-scientific engagement with nature that we no longer possess. Our modern sensibilities are not attuned to nature in the same way as our ancestors. Whether we like it or not, we were expelled from that particular

Eden when we accepted the materialist view of nature offered by science. Now when we see a rainbow we feel sentimental, rather than what we should perhaps feel, which is awestruck.

The English poet, John Keats, lamented that 'cold philosophy' would 'unweave the rainbow'. This had already happened by the early nineteenth century, the time when he was writing. But if we have lost a sense of the rainbow's mystery, science has given us something in return by opening our eyes to its complexity. Almost all the types of rainbow that you can read about in this chapter were unknown or undocumented before the seventeenth century. The rise of scientific societies during that century provided a forum where people could share their discoveries with others. Scientific journals of that era are filled with excited accounts of new phenomena, among which were reports of observations of apparently hitherto unseen rainbows.

Attempts to explain the rainbow go back to Aristotle in the fourth century B.C. But it took almost 2000 years of fruitless speculation before the correct explanation began to emerge piecemeal. Two names, in particular, are associated with the first successful explanations: René Descartes, the French philosopher and mathematician, and Isaac Newton, the English mathematician and physicist. Using the then recently discovered laws of refraction of light, Descartes was able to account for both the semicircular shape and size of the rainbow. A generation later, Newton explained its colours. In fact, it is no exaggeration to say that the rainbow has attracted the attention of the best scientific minds in every generation. The history of the scientific explanation of the rainbow reads like a 'Who's Who' of physics. You might expect that, with this degree of attention, science has done for the rainbow: explained it away, and relegated it to a paragraph or two of a seldom-read chapter in a dry tome on optics. Not a bit of it. The rainbow continues to excite interest among professional scientists. Although most aspects of the phenomenon are now well understood, there are still some unanswered questions, and from time to time, papers on the subject find their way into scientific journals.

There is yet a third reason why the rainbow is so well known. You may never have seen a halo or a corona, although both are far more common than rainbows. But you will almost certainly have seen a few rainbows. The reason for this is that most of the time we tend to confine our gaze to our immediate surroundings, and seldom raise our eyes above the horizon. To see a halo or a corona you have look up in the direction of the Sun, not an inviting prospect for the eye. We usually avoid looking at this part of the sky. But a rainbow always rises from the ground, on the opposite side of the sky from the Sun, which is why it is so much more likely that you will notice it with a casual glance than the other arcs.

In any case, a rainbow frequently announces itself with something of a fanfare: it often appears in a gloomy sky that is suddenly bathed in sunlight at the passing of a storm. The brightening alone is enough to make us glance skyward, and, as a bonus, we are sometimes greeted by a huge, bright rainbow.

One way or another, the rainbow has always formed part of our conception of nature, like clouds, or the colour of the sky. The upshot is that most people believe that they know what a rainbow is, though when pressed for details they become less certain.

5.2 How to recognise a rainbow

You probably know this much already: a rainbow is an arc of light of many colours that is sometimes seen when the Sun shines on rain. But if this is all you know, you may have seen coloured arcs in the sky that look like rainbows, such as circumzenithal arcs (see section 7.7) or coronae (see section 6.1), and thought that you had seen a rainbow.

So what distinguishes a rainbow from other arcs? What are the characteristic features of a typical rainbow?

- There are often two bows, parallel to one another. The inner one is known as the primary bow, and the outer one as the secondary bow.
- The secondary bow is always broader and much less bright than the primary bow; in fact, it is not always visible.
- The portion of sky enclosed by the primary bow is often noticeably brighter than that between the primary and secondary bows.
- The sky between the primary and secondary bow is often noticeably darker than the rest of the sky. This is sometimes called Alexander's dark band.
- All rainbows, except those formed in a fog, are composed of several parallel coloured bands.
- The outside edge of the primary bow is red and the inner one is blue or violet.
- The outer edge of the primary bow is much more clearly defined than the inner one.
- Colours in the secondary bow are the same as those in the primary bow, though they are far less bright, and their order is reversed.

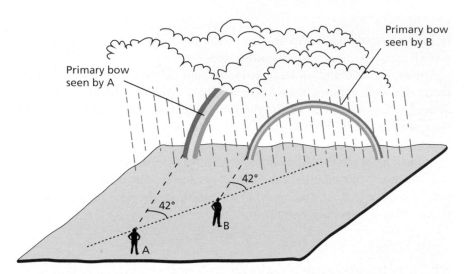

Figure 5.1 Personal rainbows. Each of us sees our own rainbow, though the angular size of the bows is always the same: 42°. Physically, however, the rainbow seen by A is larger and wider than the one seen by B because A is further from the rain in which the bow is formed. Observer A can't see a complete bow because, from her position, the cloud from which the rain is falling is less than 42° above the ground.

- Both the primary and the secondary bow lie on circular arcs whose centres lie at the antisolar point.
- The amount of arc that is visible depends on the height of the Sun above the horizon.

Strictly speaking, since the arc of a rainbow is centred on the antisolar point, there are as many rainbows as there are eyes to see them: I see my rainbow and you see yours. But, when we are next to one another, our individual bows overlap more or less completely, and to all intents and purposes we both see the same bow. It is only when we are far from one another, as in figure 5.1, that we see different bows.

Since it forms around the antisolar point, a rainbow will follow you around, like your shadow. I once saw a really bright double rainbow from a train as it travelled through South London. The bow kept pace with the train, or rather with me, for all ten minutes of the journey. As it swept across

Figure 5.2 Double rainbow. Here you can see clearly the principal features of a typical rainbow. The colours of inner or primary bow are most intense towards the ground. The sequence of colours in the outer, or secondary, bow are opposite to the sequence in the primary bow and are less intense. The sky between the bows is darker and the sky within the primary bow is brighter than elsewhere. (*Photo* Pekka Parviainen)

Figure 5.3 Sunset rainbow. The rainbow is always largest when it is formed when the Sun is on the horizon. The resulting bow and the surrounding landscape is often slightly reddened as in this photograph. (*Photo* Pekka Parviainen)

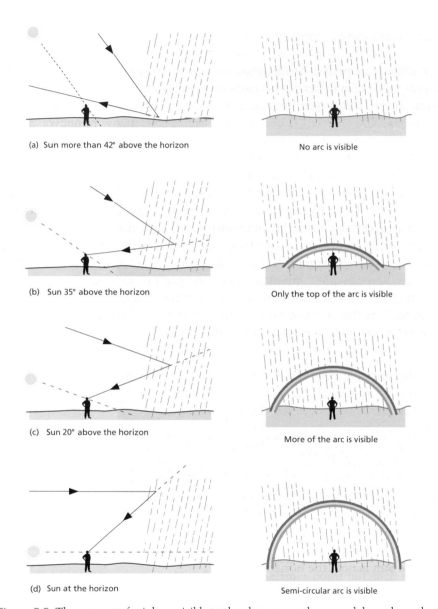

(a) Sun more than 42° above the horizon

No arc is visible

(b) Sun 35° above the horizon

Only the top of the arc is visible

(c) Sun 20° above the horizon

More of the arc is visible

(d) Sun at the horizon

Semi-circular arc is visible

Figure 5.5 The amount of rainbow visible to the observer on the ground depends on the Sun's elevation. The dashed line from the Sun through the observer's head points to the antisolar point. No bow is visible when the Sun is more than 42° above the horizon because the rainbow ray then passes over the observer's head.

These diagrams are not to scale. The observer would usually be much further from the rain than is shown here. Although a rainbow is due to a vast number of rainbow rays, only one rainbow ray is shown here to keep the diagrams as simple as possible. In any case, the rainbow rays responsible for a rainbow all enter the eye from the same direction (shown as a dotted line extending into the rain away from the observer).

local park to look for a rainbow I check the length of my shadow: if it is equal to my height, or shorter, I don't bother going.

If the Sun hugged the horizon throughout the day you would be able to see a rainbow from dawn to dusk, as long as it was raining on the opposite side of the sky. But since on most days, at least in the tropics and at mid latitudes, the Sun climbs well above the point at which it can form a rainbow, you can expect to see bows in the morning and in the afternoon far more often than at midday. Figure 5.5 will give you an idea of the effect of the Sun's height on the size of the resulting arc.

The best time to look for a rainbow is late in the afternoon as the western sky clears during a short-lived shower. Such showers favour rainbows because they tend to be confined to one part of the sky, and this makes it much more likely that the rain will be directly illuminated by sunlight. Short-lived showers depend on the solar energy, and so tend to occur late in the afternoon, which is why rainbows are seen more frequently at the end of the day than at other times.

Don't bother looking for a rainbow if the sky is completely overcast, even if it is raining: such skies often stay cloudy all day.

5.4 Supernumerary bows

Sometimes you will notice several narrow arcs, alternately pale pink and pale green, on the inside of the primary bow. These additional arcs are known as supernumerary bows, and are usually confined to the highest part of a rainbow.

Supernumerary bows are seen sufficiently often, and add so much to the beauty of a rainbow, that it is surprising that they seem not to have been widely noticed before the eighteenth century. They are a perfect illustration of the way in which science has made us more aware of nature. Supernumerary bows came to public attention through the correspondence printed in the journals of the scientific societies of the day. The following extract from a series of letters written to the Royal Society in 1721 makes clear that the writer believed that he had discovered something hitherto unseen.

> When the primary rainbow has been very vivid, I have observed in it, more than once, a second series of colours within, contiguous to the first but far weaker, and sometimes a faint appearance of a third. These increase the rainbow to a breadth much exceeding what has hitherto been determined by calculation. I remember, I had once an opportunity of making an ingenious friend take notice of this appearance, who was surprised at it, as thinking it not to be reconciled with theory.
>
> Rev. Dr Langwith

Figure 5.6 Supernumerary bows. These narrow arcs inside the primary bow are formed when raindrops are very small and of uniform size. (*Photo* John Naylor)

In fact, these bows had been noted on and off on several occasions in previous centuries, but had not been incorporated into the descriptions of the appearance of a rainbow which were current before the eighteenth century. A famous landscape by Rubens, *The Rainbow Landscape* painted in 1636, which now hangs in the Wallace Collection in London, shows what may well be a supernumerary bow on the inside of the primary arc, in the position in which one would expect to see it. Did Rubens notice the supernumerary bow by himself, or was his attention drawn to it by someone? I have not been able to discover the answer to this question. If he did notice it for himself, he had an unusually observant eye.

The width of individual supernumerary arcs depends on the size of the drops of rain in which the rainbow is formed: the smaller the drop, the broader each of the arcs. If the average size of drops differs greatly from one part of the shower to another, the width of the supernumerary bows also varies. You will often see these arcs come and go during the lifetime of a single bow.

Supernumerary bows to the secondary bow are also possible, though have seldom been seen because they are extremely faint. The following account is by Sir David Brewster, a Scottish physicist:

On the 5th July 1828 there was seen here the most brilliant rainbow that I had ever an opportunity of witnessing. Both the outer and the inner bow were perfectly complete and equally luminous in all their parts; and they continued in this condition for a very considerable time. The peculiarity in this rainbow, which has induced me to describe it at present, has I believe never been noticed. On the outside of the outer or secondary bow, there was seen distinctly a red arch, and beyond it a very faint green one, constituting a supernumerary rainbow, analogous to those which sometimes accompany the inner bow.

D. Brewster

5.5 Circular rainbows

If we were birds, and spent our days on the wing, then every rainbow we saw might be circular. As it is, earthbound, the ground gets in the way, preventing the lower portion forming, and we never see more than the upper half. But mimic the birds, and raise yourself high above the ground, either in a

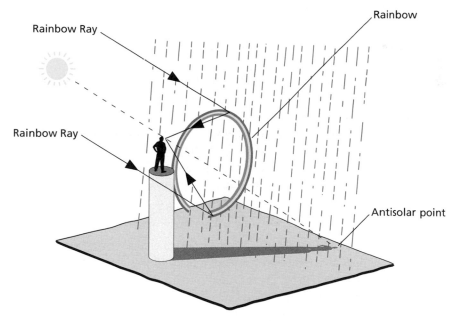

Figure 5.7 Circular rainbow. You can see a circular bow if you are able to look down on a shower of drops and are close to it. The gap at the bottom of the bow is due to the object on which the observer is standing getting in the way of the Sun's light.

Figure 5.8 Circular rainbow. A rainbow would be circular were it not that the ground prevents the lower half from being formed. However, from a tall tower or a mountain top the lower half may be visible, as this photograph shows. The curvature of the pole on the left is due to the fact that the photographs was taken with a extremely wide angle lens (focal length 16 mm). (*Photo* Claudia Hinz)

plane or balloon, or at the top of a tower, so that you can look down on a shower illuminated by sunlight, and, with luck, the rainbow you see might be completely circular.

Even if you are well above the ground, you won't see a circular rainbow unless you are close to the rain. With the Sun near the horizon, a completely circular bow will be visible only if you are slightly closer to the rain than you are above the ground. The higher the Sun, the closer you have to be to the rain.

In the following account, a rainbow changes shape as the distance between an observer aboard a ship and the rain decreases.

A rainbow of exceptional brilliance and unusual character was seen. It appeared first, 45° on the port bow, as a rainsquall was approaching, and measured about 20° in arc. As the squall reached the ship, the rainbow, which had been increasing in size and brilliance, became a complete circle, apart from the segment cut off by the hull. It seemed to rise out of the sea on both sides of the ship just forward of the bridge. The brilliance was enhanced by the bow being seen close at hand with the sea as a background.

Marine Observer's Log

Your best chance of seeing a circular bow is from an plane flying through rain, when the Sun is close to the horizon. I had a fleeting glimpse of a short segment of such a bow, which extended from the seven o'clock to ten o'clock positions on a clock dial, from an aeroplane as it was coming in to land, late one afternoon.

By the way, if you travel by plane you will sometimes see a glory. Glories are frequently mistaken for circular rainbows, though they are not at all the same phenomenon. You will find more on glories is section 6.4.

You can see a circular bow in spray from a sprinkler or a plant spray. Stand close to the drops, within a distance that is less than your height, with your back to the Sun. The resulting bow is almost a complete circle except for a segment at 6 o'clock where your shadow falls across on it.

5.6 Rainbows at sunset and sunrise

Rainbows formed at sunset or sunrise are in a class of their own. To begin with they are huge because a rainbow created when the Sun is at the horizon is a full semicircle. This is as much as can be seen of the arc of a rainbow from the ground. Occasionally these huge bows may appear entirely red because sunlight reaching the ground during the last minutes of daylight is sometimes largely stripped of its greens and blues and violets as it makes its way through the atmosphere. These missing colours won't be present in the rainbow, and what remains is a narrow red or orange arc. But you have to be quick off the mark if you want to see a red rainbow: it never lasts long because it takes only a few moments for the setting Sun to disappear below the horizon.

> There was a sharp shower for about 15 minutes just as the Sun was sinking towards a cloudless western horizon. Looking towards the east from a point about 80 ft above Wellfleet Bay, I saw an unusually brilliant primary bow at its maximum altitude, with one end in the water and the other end on the shore. As I watched, the blue, green, yellow and orange portions were quickly wiped out, the entire operation taking place in not more than 1 second. There remained a bow of a single colour, red, only slightly less brilliant than before, and in width about a quarter of that of the original bow. I turned to the west and found that the Sun had disappeared completely below the horizon, though the familiar red afterglow was still strong. No further change took place in the appearance of the rainbow for perhaps 30 seconds, when it suddenly vanished. A bow of but a single colour was visible when there was no Sun in the sky!
>
> F. Palmer

Figure 5.9 Red rainbow. A rainbow formed by the reddened rays of a setting Sun will lack the blues and greens of daytime bows because it is formed from light that lacks these colours. (*Photo* Rudiger Manig)

Rainbows can also be seen for a few moments before sunrise or after sunset, when the Sun is just out of sight below the horizon. In these circumstances the bow will be still be a semicircle, but its ends may not quite reach the ground because the raindrops closest to the ground are not directly illuminated by the Sun.

5.7 Lunar rainbows

Rainbows are formed by moonlight in exactly the same way as they are by sunlight. The only difference between lunar and solar rainbows is brightness. The full Moon is about a half a million times less bright than the Sun, and so even the brightest lunar bows usually appear colourless because the human eye can't perceive the colour of extremely faint light. Nevertheless, the characteristic rainbow colours are present in lunar bows. You can photograph these colours, and sometimes you can also see them. Very occasionally you will see a secondary bow as well.

> Yesterday evening I was fortunate enough to see a most beautiful example of a double lunar rainbow; the primary bow was very bright, quite complete and showed prismatic colours most clearly; the secondary bow was faint, but also almost complete, though the colour was barely discernible; the appearance lasted from 9.10 to 9.30 p.m. Both rainbows were apparently projected on the background of absolutely clear sky, so that at 9.20 I could clearly see the third magnitude stars θ & ψ Ursa Majoris, shining through the primary bow; this apparent absence of cloud is not infrequent with a solar rainbow, but I have never seen it before in the case of the lunar rainbow. The softness and brilliancy of the whole phenomenon was most remarkable.
>
> C.L. Brook

Lunar bows are rare because there are only a few days during a lunation, or a complete cycle of lunar phases, when the Moon is bright enough to produce a visible bow and, in any case, local rain showers usually die out after sunset. Lunar bows are often seen in Hawaii because short-lived showers are common there at all hours.

5.8 Reflection rainbows

One of the most unusual rainbows that you are ever likely to see is known as a reflection bow. It is formed by sunlight reflected by a large body of water that lies between you and the Sun. If you see a rainbow, and you have a lake or the sea at your back, look for a reflection bow, which should encircle the primary bow formed by direct sunlight, meeting it at the horizon. You will notice that its colours are arranged in the same order as those in the primary bow formed by direct sunlight, though they are not as bright because water is not a good reflector of light, except when the Sun is low in the sky. You may also see a faint secondary reflection bow.

The earliest description of a reflection bow appears in a letter written to the Royal Society by Edmund Halley and published in the *Philosophical Transactions* in May 1698. Halley reported that he saw the bow between 6 and 7 p.m. when he 'went to take the air upon the walls of Chester', a town in England. At first he saw only the primary and secondary bows, but a short while later 'with these two concentric arches there appeared a third arch, near upon as bright as the secondary iris, but coloured in the same order of

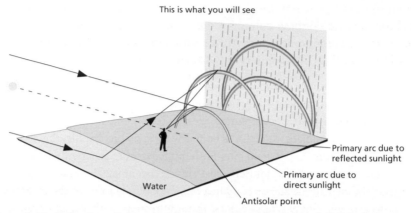

This is what you will see

Primary arc due to reflected sunlight

Primary arc due to direct sunlight

Water

Antisolar point

Figure 5.10 Reflection rainbow. Rainbow formed by reflected sunlight. Sunlight that is reflected in a body of water that lies between the observer and the Sun can produce a rainbow that will appear outside the normal primary arc.

the primary, which took its rise from the intersection of the horizon and the primary iris, and went cross the space between the two, and intersected the secondary'. Halley estimated that it lasted for about 20 minutes and concluded that the reflection bow was due to sunlight reflected by the waters of the estuary of the River Dee, which lies west of Chester.

You can see how reflected sunlight produces a reflection bow in figure 5.10. The complete arc of this bow can be formed only if the body of water in which the sun is reflected is large – as large as the rainbow to which it gives rise. You won't see a reflection bow if the Sun is reflected in a small pond.

Frequently, the only segment of a reflection bow that is visible is where it meets the primary bow: it looks like a short shaft of light springing up alongside the primary bow. These short luminous columns are the source of many of the reports of anomalous bows, i.e. of rainbow arcs for which there appears to be no explanation. You can read more about such bows in section 5.16.

Reflection bows are often seen in the Scottish Highlands, which have an ideal combination of low Sun, much rain and many large bodies of water. In 1841 Percival Frost, FRS, claimed to have seen eight bows simultaneously while walking in Scotland:

> viz., primary and secondary ordinary bows, primary and secondary bows by the reflected Sun; primary and secondary bows formed by light from the real Sun reflected from the water after leaving certain drops; primary and secondary formed by light from the Sun reflected at the water, and, after leaving other drops, again reflected at the water.
>
> P. Frost

5.9 Reflected rainbows

Whenever you see a rainbow over a large body of water, look for a reflection of a rainbow in its surface. Contrary to what you might expect, this is not a reflection of the rainbow that you see in the sky, but of one that is not directly visible to you. This reflected bow is produced by light from raindrops that lie further from you than those in which you see the bow in the sky. If you compare the arc of the reflected bow with that of the one in the sky you will see that their ends do not meet, showing that one is not the reflection of the other.

> A few weeks ago I had the pleasure of seeing a rainbow and its reflection, or at least a reflection of one from the same shower at the same time, in smooth water. The base of the bow in the cloud seemed but a few hundred yards from me, and the reflection evidently did not belong to it as the two bases did not correspond, the reflected bow lying inside the other, the red of the one commencing where the violet rays of the other disappeared.
>
> G. Dawson

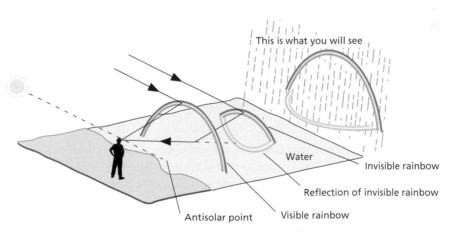

Figure 5.11 The reflection of a rainbow can be seen in the surface of a body of water. The rainbow itself is not directly visible since the rainbow rays from it can't enter the eye, except by being reflected. At the same time it may be possible to see a second bow by direct sunlight. The reflection of the invisible bow will appear smaller because it lies further away. The observer sees both bows projected against the background of water and sky.

5.10 Spray bows

On sunny days I always look for a bow when I am near a waterfall, or a fountain, or a garden sprinkler, anywhere where there is spray. I sometimes amuse myself by making spray bows with a plant spray. I try to get close enough to see a circular bow. Unlike rainbows, you can see spray bows at any time of the day, whatever the Sun's height above the horizon, as long as you can get close to the spray.

I also look for spray bows when walking along the seashore on a windy day, when the shore is being pounded by breakers. The angular radius of a bow formed in seawater spray is approximately 1° less than that formed in fresh water because sea water is slightly more refractive than fresh water. The difference can be noticed only if the spray lies between you and the rain so that both bows meet end to end.

The spray of a waterfall should be a good place to look for a lunar bow around full Moon.

5.11 Fog bows

The colours in a bow, and its size, are affected by the size of the raindrops in which it is formed. When the diameter of the drops is less than 0.1 mm, about the thickness of a sheet of paper, the resulting bow will be almost colourless. Drops as small as this are found in fogs and clouds, which is why these bows are known variously as fog bows, mist bows or cloud bows. They are also known as white bows.

These colourless bows are usually seen in a bank of fog as it clears allowing the Sun to shine on it. To see the fog bow, the Sun must be at your back. If you are close enough so that your shadow falls on the fog, you may also see a glory (see section 6.4).

> the waters of Wellfleet Bay . . . were placid at the time and covered with a blanket of thin mist some 20 ft in thickness. The Sun, still tinged with its sunrise colour was about 10° above a cloudless horizon. I waded in until the water was above my waist and then happened to glance towards the west. The appearance of the fog-bow which I then saw was striking. It was brilliant white against a background of grey mist and looked like a 'child's sized' bow made of absorbent cotton with ends dipping into the water. The overlapping of the colours was so complete that the bow appeared to be white up to the edges.
>
> F. Palmer

Figure 5.12 Fog bow. As the name suggests, fog bows are sometimes seen in fog. Fog bows are usually colourless and are broader than rainbows. (*Top photo* Jos Widdershoven; *bottom photo* Claudia Hinz)

The absence of colour in a fog bow is due to an overlap of the red and blue bands, which occurs when drops are very small. The overlap is not always complete, and sometimes a fog bow sports a reddish fringe. Although the diameter of a fog bow is always less than that of a rainbow, its arc can be up to three times as broad. Lunar fog bows have also been seen, though rarely.

Very small raindrops do not freeze at $0\,^{\circ}$C, and so fog bows, unlike rainbows which require larger drops for their formation, are possible at temperatures below freezing.

5.12 Rainbow wheels

Shadows cast by clouds that lie between the Sun and a rainbow can prevent sections of a bow from forming. The resulting rainbow will be chopped up into segments of unequal length, arranged along a circular arc, resembling the spokes of a wheel.

> [at about 6.30 p.m.] . . . a very fine and complete bow was seen, partly against the sky, partly against the hills which bound on the south-east the high road from Alpnach. At intervals several radial streaks were observed both interior and exterior to the rainbow. They shifted position and magnitude rapidly. It was noticeable that to each streak corresponded a patch of sunshine on the hills behind, each streak pointing to a patch and moving with it.
>
> S.P. Thompson

The wedge-shaped radial streaks in this description were shafts of sunlight streaming through gaps between clouds on the opposite side of the sky from the rainbow. Because of perspective, shafts of sunlight appear to converge at the antisolar point, the same point on which the rainbow is centred, and this makes them appear like wedges of light radiating from the centre of the bow.

You can also see short segments of a rainbow in distant showers, or showers that are physically small. A complete rainbow can't be seen if the apparent size of the shower is smaller than the bow. Mark Twain describes seeing these in Hawaii:

> What the sailors call 'rain dogs' – little patches of rainbows – are often seen drifting about the heavens in these latitudes, like stained cathedral windows.
>
> Mark Twain

5.13 Horizontal rainbows

We expect to have to look up at the sky to see a rainbow. However, rainbows are sometimes seen on horizontal surfaces, such as grass lawns and lakes, when small drops either rest upon the surface, or are suspended a short distance above it. A horizontal rainbow is produced in the same way as a rainbow in the sky: a combination of reflection and refraction.

A striking feature of this type of bow is its shape. This depends on the height of the Sun above the horizon. When the Sun is directly overhead, the bow will be circular, though no one has ever reported seeing a circular horizontal bow. But, when the Sun is close to the horizon, the bow resembles an inverted arch with its apex at your feet and its ends extending away into the distance on either side of you. In geometry this shape is known as a hyperbola.

If you are to see a horizontal bow, you are most likely to do so early in the morning during autumn as a radiation fog clears. Grass lawns that are covered in a fine spider's web, or gossamer, make ideal surfaces. Drops from the fog are deposited on the fibres of gossamer. As the fog clears, and the drops are illuminated by a low Sun, you may notice the faint hyperbolic arc of a horizontal rainbow. These bows are often called dew bows, though this is a misnomer since the drops on the webs are not dew, they come from fog.

> The place, the conditions and the time were as follows. On Friday, November 5th, at Tadmorton Heath golf course, situated on the high ground near Banbury, there was a thick fog which persisted until about 11.30 a.m. in spite of a Sun which seemed to be struggling to penetrate it. The ground, more particularly noticeable on the fairways, was literally covered with a dense carpet of cobwebs which, being saturated either with particles of moisture or the remains of a ground frost – I do not know which – presented an appearance suggestive of snow. At 11.45, when the fog had suddenly dispersed and a brilliant Sun was almost due south, my companion and I came to a fairway running directly north and, upon stepping on to the mown and heavily cobwebbed portion, were simultaneously arrested by the extraordinary spectacle of a perfect rainbow flat on the grass. It started at our feet and ran in an elliptical form, definitely not a circle, to the extremities of both sides of the fairway and continued in front of us until the lightly mown turf of the green was reached. It was faint compared with the brilliance of an ordinary rainbow in the sky, but in all other respects was identical, and, culminating as it did at our feet, it added appreciably to the recognised

Figure 5.14 Rainbow formed in a narrow beam of light. As the beam swings away from the observer from A to C, the section of rainbow visible in the beam appears to move along the beam towards the source. The red end of the bow is closer to the source than the blue end, though colours are difficult to make out because, despite appearances, a searchlight beam is not very bright.

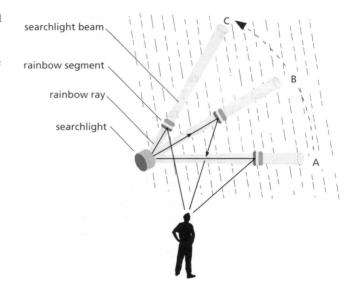

5.15 Eclipse rainbows

On at least one occasion, a rainbow may have been formed by light from the Sun's corona (see section 10.6) during a total eclipse. E.W. Maunder, reporting on an eclipse seen in Mauritius on 18 May 1901, wrote that a rainbow was seen at the height of the eclipse and that all the observers present agreed 'that it tapered at the two ends, and that [one of the observers] had been much impressed by bright lines running through it, particularly a bright pink line. Could this have been the C line of hydrogen from a large prominence?' A prominence is a mass of hot gas, mainly hydrogen, ejected from the Sun's surface. Being hot, it glows with a colour that is characteristic of the gas of which it is composed: hot hydrogen glows with a pinkish light.

5.16 Anomalous rainbows

Although the scientific theory of the rainbow is more or less complete, it has to be applied cautiously to natural rainbows. The idealised assumptions on which the theory is based seldom, if ever, exactly match the conditions in which rainbows are formed in nature. Hence, very occasionally, you may see, or hear about, a rainbow with features that appear to defy explanation. Such bows are known as anomalous rainbows.

Broadly speaking, anomalous rainbows fall into three broad categories.

- Most reports of unusual rainbows are the result of a lack of knowledge of the full range of optical phenomena in nature. Circumzenithal arcs, circumhorizontal arcs (section 7.7), and glories (section 6.4), are all multicoloured arcs that are frequently mistaken for rainbows by inexperienced observers.

- A second source of anomalies is the result of careless observers who report incorrectly the details of what they see. Over the years this has given rise to a pernicious body of phenomena that cannot be explained, or easily dismissed because the observer is no longer around to establish the correct details of what was really seen.

- Finally, from time to time, genuine anomalies are seen that challenge our understanding of rainbows. In these cases, it is usually a matter of accounting for small, though important, details of the circumstances that would be necessary for such an unusual feature to occur.

Reflection bows, particularly those limited to a short vertical shaft next to the foot of the primary bow, are probably the type of bow most often mistaken for an anomalous bow.

The major difficulty with genuine anomalies is not that rainbow theory cannot explain them; the problem is usually how to account for the circumstances that would give rise to such a bow. R.W. Wood, who was an eminent authority on optics, claimed to have seen a rainbow of variable curvature 'the left half appearing to be pulled out into an arc of greater curvature'. Similar bows have been seen on other occasions. It is known that the diameter of a rainbow depends on the size of the raindrops in which it is formed. So the change in curvature could be explained by assuming that the average diameter of drops in which one part of the bow was seen was not the same as that in which the other part was seen. The difficulty here is to explain how drops can be significantly larger in one part of a shower than in another.

Another unusual observation is that the bow appears to vibrate, with colours fading and reappearing in rapid succession. This has been seen very rarely and may be due to the effect of the shock wave of a lightning stroke on the shape of the raindrops. Recent investigations into the effect that the shape of raindrops have on the appearance of rainbows seem to support the view that a shock wave from a lightning stroke may cause larger raindrops to oscillate. As a drop oscillates, its cross-section changes. A rainbow ray emerges in the usual direction only when the cross-section of a drop is spherical, or nearly so. So, as drops oscillate, portions of the bow may periodically fade from view. However, many details remain to be worked out and a complete explanation for these oscillating bows has yet to be given.

5.17 Explaining the rainbow

A complete explanation of the rainbow, couched in the mathematics of the wave theory of light, is not for the fainthearted. Fortunately, it is possible to account for most of its principal features in terms of reflection and refraction of light, without invoking the wave nature of light.

People realised long ago that, since rainbows are always seen on the opposite side of the sky from the Sun, they are due to reflected sunlight. But if raindrops reflected light as efficiently as mirrors, rainbows would be blindingly bright. They would be amazing sights, though we wouldn't be able to look at them directly without endangering our eyesight. As it is, almost all the light that enters a raindrop passes straight through it because water is a transparent medium, so rainbows are rather faint. A rainbow is formed by the tiny amount of light reflected from the inside surface of each one of the multitude of drops that make up a rain shower.

Reflection alone can't explain why we see a circular arc of many colours when drops of water are illuminated. Refraction is also necessary. Refraction is the change in direction that occurs when light passes from one medium to another. Figure 5.15 shows the path of narrow bundles of light rays through a drop with a circular cross-section. Notice that two bundles of rays, in which rays are parallel to one another on entering the drop, diverge when they emerge from it. These diverging rays are responsible for the diffuse brightness that is sometimes noticeable within the arc of the primary bow. The rainbow itself is formed by a particular bundle of rays that emerge from the drop more or less parallel to one another. They make up what can be called the 'rainbow ray'.

The rainbow ray has a unique property: it is made up from rays that are deviated least from their original path. This angle of minimum deviation, as it is known, depends on the wavelength of light. For red light in water it is approximately 138°, and for violet light, 140°. In other words, the angle between light from the Sun and the rainbow ray responsible for the red band on the outside of the primary bow is about 42°, and for the inner violet band it is about 40°. The width of the bow is thus roughly 2°, about the width of your thumb held at arm's length, though natural bows are slightly broader than this for reasons given below.

The path of the rainbow ray represents a limit to the deviation, or change

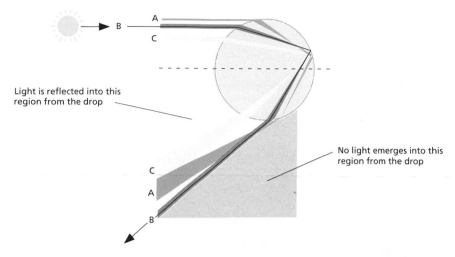

Light is reflected into this
region from the drop

No light emerges into this
region from the drop

Figure 5.15 The rainbow ray. The diagram shows the path of some light rays through a
drop with a circular cross-section. B is the rainbow ray. See text for a full explanation.

in direction, that light undergoes when reflected and refracted by a raindrop:
rays can be deviated through larger angles, but they can't be deviated
through a smaller one. Hence no light emerges from the drop into the region
beyond the rainbow ray, which is why the sky between the primary and secondary bows sometimes appears distinctly darker than the rest of the sky.

More importantly, the absence of light in the region beyond the rainbow
ray allows you to see the rainbow ray itself as a coloured fringe to the light
reflected in your direction by each drop. Without the cut-off due to
minimum deviation, a rainbow would not be a coloured arc. It would merely
be a diffuse brightness around the antisolar point, rather like a *heiligenschein*, which most people would probably never notice.

You may find all this easier to understand if you carry out the following
simple experiment. Fill a test tube with water – if you don't have a test tube
a clear glass rod works just as well – stand with your back to the Sun, and hold
the tube vertically at arm's length, within the shadow of your head. Keeping
your arm outstretched, slowly swing it out from the shadow, ensuring that
the Sun's reflection is always visible in the tube. At some point the Sun's
image becomes strongly coloured: you are seeing the rainbow ray. Which
colour do you see first: red or violet? Can you see more than one colour at a
time? Through what angle have you moved your arm? See if you can identify
the rainbow rays responsible for the primary and for the secondary arcs. Is the
dark band between the primary and secondary arcs noticeable? If you have

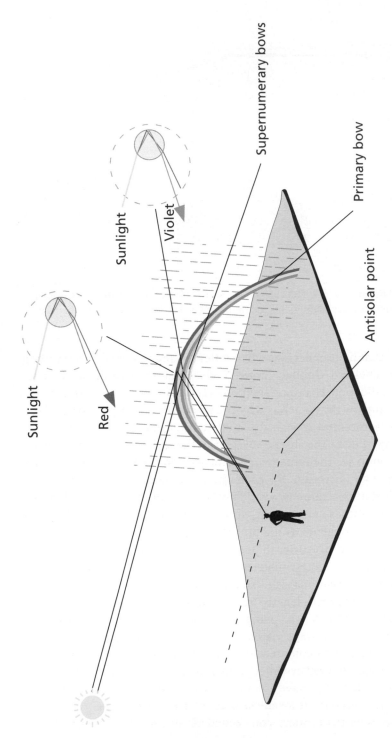

Figure 5.16 The primary arc of a rainbow is due to sunlight being reflected once within a raindrop. The outer edge of the arc is red, the inner edge blue or violet. Sometimes you will also notice a series of narrow coloured bands along the inner edge near the top of the arc. These are known as supernumerary bows.

Sunlight

Red

Sunlight

Violet

Supernumerary bows

Primary bow

Antisolar point

tubes of different diameters you can also check if the brightness of the rainbow is due to size of the raindrops in which it is formed.

The most striking feature of a rainbow is its colours. These are due principally to refraction, which separates sunlight into its constituent colours as it passes though a raindrop. The resulting spectrum is not pure, and there are several reasons for this.

In the first place, sunlight is not perfectly parallel and so adjacent colours, because of refraction within a drop, overlap one another slightly. And rainbow rays of different colours diverge slightly on leaving a drop, which increases the amount of overlap. Colours in a pure spectrum, formed when a very narrow, parallel beam of sunlight is viewed through a glass prism, are well separated, and consequently are far more vivid than those in a bow.

Secondly, there are two mechanisms at work within a drop that cause colour: refraction and interference. Interference in drops with a diameter greater than about 1 mm is insignificant, and the dominant cause of colour in larger drops is refraction. But in very small drops the effect of interference between rays close to the rainbow ray becomes increasingly noticeable. This causes some colours, such as red and blue, to disappear and creates a succession of clearly separated bows, supernumerary bows (see section 5.4), just inside the primary arc. Depending on the range of drop sizes within a rain shower, the occurrence and relative intensity of various colours differ noticeably from one bow to another. Colours may also change during the lifetime of a particular bow if a large proportion of drops become smaller through evaporation.

Finally, drops are transparent, and so the colour and brightness of a bow is influenced by the colour and brightness of the background against which it is seen. Although the brightness of a rainbow increases with the depth of the shower in which it is formed, the shower itself also becomes brighter. And since the eye judges how bright a thing is by comparing it with its surroundings, an increase in the brightness of a shower makes a bow appear less bright than it really is.

Depending on the depth of the rain in which they are formed, the brightness of the primary and secondary bows, the space between them, and the space within the primary bow varies from one rainbow to another. In a shallow shower, the primary bow and the space it encloses will be faint, and the dark space hardly noticeable. The brightest bows are seen in light showers when the background is either a clear, blue sky or dull, uniformly grey clouds.

So much for what happens within a single drop. How does this explain the arc of a rainbow?

Have you ever have noticed, when looking at dew on a lawn early in the morning, that to see more than one colour in a particular dewdrop you have to move your head? The same is true of raindrops: your eye can only intercept a ray of a single colour from each drop. Other colours come from neighbouring drops. The bow itself is a mosaic of different colours that can be seen in its entirety only in a broad expanse of rain.

The bow's shape is due to the fact that the only rainbow rays that can enter your eye come from drops that all lie at the same angular distance from the antisolar point. This circle has an angular radius equal to the angle that rainbow rays emerging from each drop make with light from the Sun: 42° for red light, 40° for violet.

The resulting rainbow does not lie at a particular distance from you, though it appears to do so. It looks as if it's a flat arc, enclosing a region that is brighter than the surrounding sky, but it's actually three dimensional: a cone of light, fringed with colour, which stretches away from you into the rain. You can never see the cone itself because your eye is at its apex, and its base lies within the shower, so you are unaware of the rainbow's depth, of its three-dimensional nature. The arc is merely the most eye-catching aspect of this cone of light. We see it projected against the sky, which makes it appear flat.

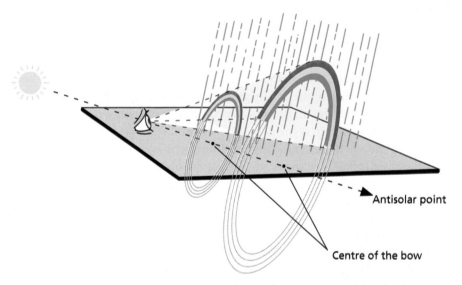

Antisolar point

Centre of the bow

Figure 5.17 Rainbow cone. Although a rainbow appears to be a semicircular arc it is really a cone of light that stretches away from the observer and into the shower of drops in which it is seen. The lower half of the cone is not seen because ground gets in the way: there are no drops below ground level, and no sunlight to illuminate them.

The fact that the arc lies on the surface of this cone imposes a limit on the greatest distance at which you can see a complete rainbow in a shower. The apex of the bow is formed in drops that lie in the direction of the clouds from which they are falling. Hence it is the height of the clouds that limit how far away you can see a complete bow. For example, if the clouds are 300 m above the ground, and the Sun is 15° above the horizon, then the greatest distance at which you will see a complete bow is approximately 600 m. In this case, if you are more than 600 m from the rain, then you won't see the top of the rainbow, though you should see its ends.

The secondary bow is formed by two internal reflections of the rainbow ray. As with the primary bow, only a fraction of the beam is reflected each time. The amount of light remaining after two internal reflections is barely enough to be visible, which is why the secondary bow is so much fainter than the primary bow. Indeed, it is often not visible. At the same time, the colours in the secondary bow are in reverse order to those in the primary bow. Figure 5.18 shows that the secondary bow is due to drops that are further away from the ground than those in which the primary bow is formed. The angular radius of the secondary bow is approximately 50° and it is approximately 3.5° wide. The separation between the primary and secondary bows, i.e. Alexander's dark band, is about 9°.

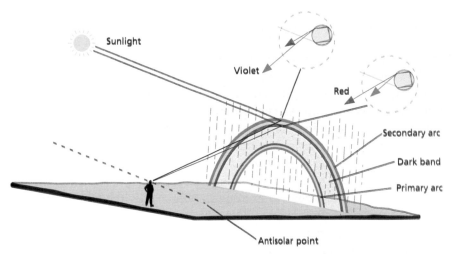

Figure 5.18 The secondary arc. Sometimes a second arc is visible outside the primary one. The drops in which this secondary arc is seen are further from the antisolar point than those in which the primary arc is seen. The gap between the arcs is often noticeably darker than the sky outside the secondary arc and inside the primary arc. Notice that the order of colours seen in the secondary arc are the reverse of those seen in the primary arc.

The principal weakness with the explanations that you have just read is that they are based on the passage of parallel beams of light through a idealised spherical raindrop. Although this explains the major features of the rainbow, there are many other features observed in natural rainbows that it cannot explain. The reason for this is that the theory does not take into account the shape of real raindrops, ignores the brightness and colour of background against which bows are actually seen, and assumes that the Sun is a point source of white light. Of these, the shape of the drop has the greatest effect on the appearance of the bow.

Most rain showers are composed of a range of drop sizes, from diameters as small as 0.2 mm to as large as 5 mm. Although the smallest drops are always spherical, larger ones become flattened as they fall, and so their vertical cross-section is not circular, though their horizontal one may be. Such a drop resembles a miniature bun, rounded on top and flattened underneath; raindrops are never pear-shaped. The rainbow ray that emerges from the vertical cross-section of a flattened drop undergoes greater deviation than the one that emerges from its horizontal cross-section. This should lead to a bow of varying curvature, with a vertical radius that is less than the horizontal one. However, smaller drops within the rain shower are not flattened to the same degree as the larger ones, and they give rise to a bow of constant curvature. The rainbow rays from the vertical sections of larger drops fall inside this bow and contribute to the general brightness inside the arc.

Distortion of raindrops also explains why the brightness of a rainbow is usually least along the apex of the arc. The proportion of drops that contribute rainbow rays is less there than it is closer to the ground, where drops of all sizes contribute rainbow rays to the vertical portions of the bow. Rainbow rays that form the vertical portion of a rainbow are more or less parallel to the ground. Since even the largest drops can have a circular horizontal cross-section, their rainbow rays can emerge in the same direction as those from smaller drops, giving rise to a brighter bow. Of course, this can be noticed only when a bow has a vertical portion, which occurs only when the Sun is close to the horizon.

The range of drop sizes within a rain shower makes it difficult to explain why we see supernumerary bows. These bows become more widely spaced, and are therefore more noticeable, as drop size decreases. In a shower composed of drops of many sizes, however, supernumerary bows due to one drop size should overlap those produced by other sizes. The net effect is that supernumerary bows should not be seen in most showers, they should be visible only when the drops are all of a similar size, for example in drizzle. Nevertheless, supernumerary bows are sometimes seen in heavy showers of rain, which are known to contain drops of different sizes.

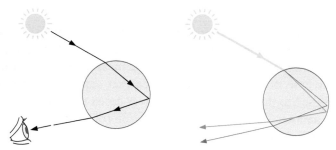

(a) Primary arc: refracted twice, reflected once.

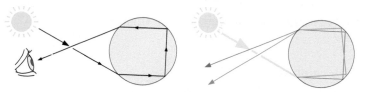

(b) Secondary arc: refracted twice, reflected twice.

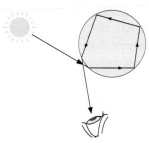

(c) Tertiary arc: refracted twice, reflected three times.

(d)   (e)

Figure 5.19 Diagrams (a), (b) and (c) show the path of rays through an idealised raindrop that has a circular cross-section. In practice, as raindrops grow larger, they become flattened as they fall to the ground. The rainbow ray undergoes greater deviation in the flattened raindrop (diagram e) than it does in the round raindrop (diagram d). Those parts of a rainbow formed in flattened drops therefore have a smaller diameter than the rainbow formed in spherical drops.

A possible explanation is that supernumerary bows due to smaller drops are noticeable because larger drops flatten slightly as they fall, and contribute their rainbow rays to the general brightness inside the primary bow. If this is the case, it also explains the fact that supernumerary bows are usually confined to the top of the bow. Lower down the arc supernumerary bows due to smaller drops in the vertical sections of a bow are overwhelmed by the light from larger drops.

The width of a natural rainbow is about half a degree greater than that predicted by theory because the Sun is not a point source and the rainbow ray is not precisely parallel light. Raindrops illuminated by a diverging beam not only produce a wider bow; the colours are less well separated, and therefore not spectrally pure, because they overlap slightly.

5.18 Tertiary rainbows

Rainbows of a higher order than the secondary bow are possible in theory. Three internal reflections of the rainbow ray would produce a tertiary bow, and four internal reflections would give a quaternary bow. In practice, these bows are much too faint to be seen against the brightness of the sky, through they can be observed in the laboratory. Figure 5.19 shows that the rainbow ray responsible for the tertiary bow emerges from a drop on the opposite side to those which cause the primary and secondary bows. This means that if you wanted to see a tertiary bow you would have to look at the sky in the vicinity of the Sun. Theory predicts that it should be a circular arc of diameter approximately 40° centred on the Sun with its red edge on the outside.

The possibility of a tertiary bow has been known ever since Edmund Halley calculated its position. Generations of experienced observers have taken up the challenge and have looked for it, though without apparent success. Ice halos and circumzenithal arcs are frequently mistaken for tertiary bows. If you want to take up the challenge of seeing a tertiary bow you should arrange for the Sun and the sky around it, right up to the point where this bow should appear, to be blocked off, say by a building. No one has ever reported seeing a quaternary bow in nature.

The most recent sighting of the elusive tertiary bow was made by D.E. Pegley, an experienced and reliable observer. As you can tell from his account of the event, although he was uncertain whether he had seen the bow, it is clear that he did not see an ice halo (section 7.1). So, what did he see?

Whilst in Nairobi recently I had the good fortune to see a tertiary rainbow. On 21 May 1986 at 17.55 a new shower cloud had just started to rain out over my hotel in dense curtains of medium-sized drops brilliantly lit by the low Sun. From the balcony of my fourth floor room I could see not only a bright primary, accompanied by a moderate secondary, but also a weak bow in the direction of the Sun, which was conveniently shielded by the side of the building. The bow was scintillating but distinct for two or three minutes. It was about the same size as the primary bow, with red on the outside and green on the inside . . . was it my imagination, or have other readers seen a tertiary rainbow . . .?

D.E. Pegley

5.19 Polarised rainbows

Light from a rainbow is polarised because it is formed by reflected light. The rainbow ray is particularly strongly polarised (see section 1.6) because the angle at which it is reflected within the drop is very close to the Brewster angle (see section 1.8) for water. The high degree of polarisation can be confirmed by looking at a rainbow through a polarising filter: any portion of the bow that can be seen through the filter vanishes completely when the filter is rotated through a quarter turn.

If you hold the filter against your eye so that you can see the whole bow through it and then rotate the filter, you will notice that, in one position, the apex of the bow is visible but not the foot, while a quarter turn brings the foot into view as the apex vanishes. This suggests that these parts of the bow are polarised perpendicularly to one another. This is to be expected because the rainbow rays responsible for the apex of the bow pass through the drops in a vertical plane while those responsible for the ends of the bow pass through the drops in a horizontal plane. Light polarised by reflection is polarised in a direction that is perpendicular to the reflecting surface. Hence a rainbow is polarised in a direction that is tangential to the bow. In other words, the plane of polarisation of a rainbow follows the curve of the bow. Both primary and secondary bows are tangentially polarised.

5.20 Are rainbows real?

Since no one else can see the same rainbow as you, can it be real? The light that forms the arc is certainly there, but what of the arc itself? It's such a palpable thing that it is difficult to accept that you are not looking at something that really is in the sky, like a cloud, and that you are in fact creating a bright arc by intercepting rays of coloured light with your eyes.

You might argue that everything that we see is due to rays intercepted by the eye. And, if so, isn't everything as real or unreal as we suppose rainbows to be? Well no: the arc, unlike a cloud, doesn't exist independently of the observer. It really is in the eye of the beholder.

This point may be easier to understand if you think about a simpler phenomenon than a rainbow. Take a glitter path, the narrow, flickering, swathe of light that we see when looking towards the Sun over choppy water. It is made up of multiple reflections of the Sun. The crest of every wave over the entire surface of the water reflects a distorted image of the Sun, but you see only those from crests that reflect light directly towards you. The law of reflection prevents you from seeing any reflections except those within a narrow strip either side of a line directly between you and the Sun.

It is the same with the rainbow. Just like the glitter path, it is not located in any particular place. Its position depends on where you are when you see it. Every sunlit raindrop reflects rainbow rays at the same angle. Those that miss your eye are not seen by you. And the law of reflection dictates that those that you do see lie within a narrow circular arc.

Another consequence of the rainbow's dependence on the observer is this. If you see a rainbow and begin to walk towards the point where it touches the ground, you discover, assuming the bow obliges by continuing to shine, that the end you are trying to reach recedes from you as you approach. This happens because the bow is not physically fixed in space: it is a mosaic of colours centred on your eye. If you move, so does the bow.

Don't wait for a natural rainbow to discover this. It's easier to check this out by making a bow with a plant spray because you can create the bow within arm's length. On a sunny day, get someone to work the spray as you walk slowly towards the bow. You will find that, as you get closer, the bow eventually splits in two, and then vanishes as you pass through the droplets.

You see two bows when you are less than about a metre from the spray because each eye sees a rainbow in a different set of drops. If you close one eye, its bow vanishes. But even the double bows disappear as you get closer to them because they become physically smaller. The band of the primary bow is approximately $\frac{1}{30}$ radian broad. At 300 m it will be 10 metres wide, which is why at a distance quite large objects may appear to be completely immersed within a bow. At arm's length, however, the bow narrows to about 2 cm and, within a few centimetres of the eye, it will be only a few millimetres wide. But you won't be able to see it because, when you're enveloped by the spray, your head prevents sunlight reaching the drops, and the bow vanishes.

What people probably have in mind when they speak about getting to the end of the rainbow is that they think they will be immersed in a multi-coloured pool of light, the same colours that can be seen when you see the bow at a distance. Unfortunately, as the experiment with the spray bow shows, this is not possible: as we approach the bow, it just gets smaller and smaller, finally vanishing just as we reach it. It strikes me that there is a kind of fairy-tale morality about this. If your motive for reaching the end of the bow is to get your hands on that legendary pot of gold, then it is fitting, in a fairy-tale sort of way, that not only is there no pot of gold there, but the rainbow itself vanishes just as you get close enough to reach for it.

The closest I have ever been to a natural rainbow was about 20 metres. It formed in front of a row of trees on the other side of a playing field late one wet and blustery autumn afternoon. The left-hand end of the bow touched the ground and was very faint because the rain in which it was formed was only a few metres deep. The thrill of it was that we are seldom as close to something that is quite as otherworldly as a rainbow.

5.21 Notes for rainbow observers

A rainbow offers quite a challenge to your powers of observation. There is so much to see, and, usually, so little time to see it. Beyond the fact that it is a semicircular arc composed of a succession of parallel coloured bands, it is not easy to notice its finer points without some prior knowledge of what to look for. The questions below, which are in no particular order, may be of help to you.

1 Make a note of the time, date and place. In my experience, afternoon bows outnumber morning and noontime ones by a large margin. If possible, also make a note of how long it lasts.
2 What is the nature of the shower in which the bow is seen: drizzle, light rain, heavy downpour? Can you estimate how far away the rain is from you?
3 Is it possible to tell how far away the bow is? Make an estimate.
4 Is the arc of the primary bow complete? If not, is this because the rain is at a great distance from you or is it because the rain is confined to one part of the sky?
5 Is a secondary bow visible?

6   How apparent is the dark band between the primary and secondary bow, and how bright is the arc enclosed by the primary bow? Is any colour noticeable in either the dark band or within the area enclosed by the primary bow?

7   Are supernumerary bows visible? How many are visible? More than three or four is unusual. What is the angular width of each one? Measure several together using the outstretched hand method (see Appendix page 301) and divide by the number of bows.

8   What colours are present in the various bows? Make a record of them.

9   Do the same colours continue along the entire length of the arc?

10   Are some colours more vivid than others, i.e. are some colours brighter or do they form broader bands than others do? Which ones?

11   Do the colours change during the lifetime of the bow? What are these changes?

12   Does the brightness of the bow vary along the arc? Where is it brightest? Compare the apex to the foot of the bow.

13   What is the background against which the bow is seen: cloud, blue sky, landscape?

14   Does the colour or brightness of the background appear to affect the colour and brightness of the bow? In what way?

15   If you have a polarising filter handy, examine the bow through it, and determine the direction of polarisation along different portions of its arc.

If you want to photograph a rainbow, bear in mind that a primary rainbow formed by a low Sun is some 85° across and 42° high. You can just fit this in the frame of a 35 mm film using a 17 mm wide angle lens. Under-expose to bring out the colours in the bow.

Chapter 6 | CORONAE AND GLORIES

> For in June, 1692, I saw by *reflexion* in a vessel of stagnating water three halos, crowns, or rings of colours about the Sun, like three little rainbows, concentric to his body. The colours of the first or innermost crown were blue next the Sun, red without, and white in the middle between the blue and red. Those of the second crown were purple and blue within, and pale red without, and green in the middle. And those of the third were pale blue within, and pale red without; these crowns enclosed one another immediately, so that their colours proceeded in this continual order from the Sun outward: blue, white, red; purple, blue, green, pale yellow and red; pale blue, pale red . . . like crowns appear sometimes about the Moon.
>
> Isaac Newton, *Optics*, Book 2 Part 4, G. Bell & Sons, 1931

6.1 Coronae

When a Moon that is full, or nearly full, is seen though a thin veil of altocumulus cloud, you may sometimes notice that it is surrounded by a series of more-or-less concentric, coloured rings. These are known as diffraction coronae. Corona is Latin for crown. Coronae also form around the Sun in the same circumstances, though you are unlikely to notice any colours in clouds that lie close to the Sun because of the brightness of the clouds. In any case, it's not worth running the risk of ruining your eyesight by looking for coronae around the Sun. Looking at the Sun directly can permanently damage your eyesight. You can get around this problem by looking at a reflection of the Sun in a puddle of clear water, or a windowpane. Indeed, this is how some people, previously unaware of the phenomenon, have seen it for the first time: they inadvertently notice that the Sun's reflection in the smooth surface of a body of water is surrounded by coloured rings. This is how Isaac Newton came to see the phenomenon. He was sufficiently taken by what he saw that he made careful notes of its appearance. The fact that he also made a note of the date suggests that, as far as he could tell, it was an unusual phenomenon. In fact, it is a very common one.

It is no exaggeration to say that diffraction coronae occur more often than almost any other optical effect in nature, yet there is no fan club for the phenomenon, as there is for rainbows and halos. Instead, diffraction coronae

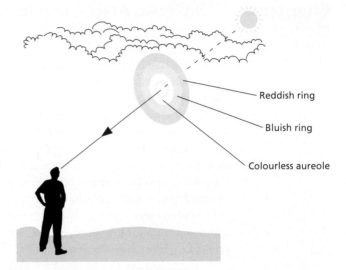

Figure 6.1 Coronae are those coloured rings than we often see when the Sun, and more especially the Moon, is seen through thin cloud. The rings are concentric, alternately bluish and reddish, centred on the disc of the Sun or Moon.

Reddish ring

Bluish ring

Colourless aureole

are often confused with rainbows or halos, and sometimes with the solar corona that is seen around the Sun when it is eclipsed by the Moon. But when you have seen a few coronae you will easily tell them apart from other optical phenomena. In the first place, diffraction coronae are multicoloured, and are always very much smaller than ice halos (see section 7.1). Secondly, they are formed within the atmosphere, and are not part of the Sun.

Coronae can be formed in any cloud that is made up of tiny droplets of similar size. Hence they are frequently seen in altocumulus, cirrocumulus, and in fog. All of these are composed of tiny droplets of water. Coronae have also been seen in clouds composed of ice crystals, though far less frequently than in clouds of water drops. In all cases, the cloud must be thin enough so that you can see the Sun or Moon through it. The very best coronae, i.e. those that consist of perfectly circular coloured rings, form only in clouds composed of particles that are all of the same size. The particles can be water droplets, tiny ice crystals or pollen grains. Particularly striking coronae can be seen in lenticular clouds. These clouds, which sometimes form above mountains when moist air is forced to rise, are composed of tiny water droplets of similar size. If the particles aren't all the same size, the resulting corona will be an irregular, confused mixture of colours. Clouds that are coloured in this haphazard way are known as iridescent clouds.

In a perfectly formed corona, the inner edge of each ring is blue and the outer one red. More often than not, only the red edge of the innermost ring is apparent. The area enclosed within it, known as the aureole, is a bluish

Figure 6.2 Corona around the Moon. This appears as a series of coloured rings around the Moon when the Moon is seen through cloud. (*Photo* John Naylor)

white. This bluish aureole, without an outer red rim, is often apparent when the Sun is seen through wisps of cumulus. Sometimes a series of concentric rings are visible, the number of which depend in part on the brightness of the source of illumination. Even when the Moon is full, you won't often see more than two sets of rings around it, while around the Sun more than three or four are unusual. However, it is sometimes possible to see fragments of the fifth, sixth and even seventh ring.

The angular diameter of each ring depends on drop size and so varies from one cloud to another. The radius of the innermost ring ranges from 2° to 4°, which makes it 4 to 8 times the apparent diameter of the Sun or Moon. But since drop size varies greatly from one cloud to another, rings with radii outside this range have been observed on occasion. Colours in the second set of rings tend to be better defined than in the first one.

Drops on the edge of a cloud are evaporating, and so are sometimes significantly smaller than they are within the cloud. Coronae seen in such cloud are not circular: their radius increases towards the edge of the cloud. The resulting corona takes on the shape of a horseshoe.

A full explanation for coronae is beyond the scope of this book. Briefly, when light passes through a cloud of particles, it is diffracted by

Figure 6.4 Cloud iridescence. Sometimes clouds close to the Sun become strongly multicoloured. (*Photo* Pekka Parviainen)

are seen. These patches of light often form haphazard patterns. Occasionally, quite large sections of a cloud take on particular colours. In this situation, bands of colour may be aligned with the edge of the cloud. For example, one of the edges of the cloud might be pastel pink, with a pastel green band running parallel to it closer to the centre of the cloud.

Nacreous clouds are rarely-seen clouds that occasionally form in the stratosphere at altitudes of 20–30 km. Although little is known about them, they are thought to be a type of lenticular cloud because they appear to remain stationary above mountainous regions. Nacreous clouds that are seen at sunset, and are within a few degrees of the setting Sun, may become markedly iridescent because of the diffraction of sunlight by the cloud droplets. They are sometimes called mother-of-pearl clouds ('nacreous' means 'mother-of-pearl'). They are usually visible for several hours after sunset because of the great height at which they are formed. They have been seen from Scotland, Iceland, Norway and Alaska, though they are most frequent over Antarctica. They are usually seen in winter and it is thought that particularly low temperatures in the stratosphere may be an important factor in their formation.

6.4 Glories

Under the right circumstances, you may see multicoloured rings around your antisolar point. This is a glory, which is formed by light reflected by the droplets of water that make up clouds and fogs. This sounds rather like what happens when a rainbow is formed. However, refraction doesn't play a part in forming a glory. In the first place, cloud droplets are too small to form conventional coloured rainbows, though they can form fog bows and, indeed, if the cloud or fog bank is extensive, you may see a fog bow at the same time as a glory. Secondly, the glory is formed by light that is reflected directly back in the direction from which it came. A thorough explanation of how it is formed involves some quite advanced physics, and is beyond the scope of this book.

To see a glory, your antisolar point must fall on a cloud or bank of fog, and the Sun must be shining. Depending on how far you are from the cloud or fog, you will see a series of coloured rings surrounding the shadow of your head. You can also see glories at night around your shadow when this is cast on fog by a bright light.

You are much more likely to see a glory in mountainous regions than

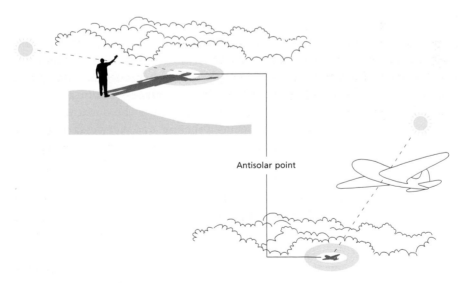

Figure 6.5 A glory is a series of circular concentric coloured bands sometimes seen around a shadow on a cloud. You might see one around your shadow if this falls on cloud or mist, or around the shadow of an aeroplane in which you are a passenger. A glory is always centred on the antisolar point.

Figure 6.6 Glory. The triangular shape is the shadow of the photographer and the glory surrounds the shadow of the photographer's head. (*Photo* John Naylor)

you are in lowlands. In fact, during the nineteenth century, glories became associated with particular mountain sites. Glories were seen so frequently in the Hartz Mountains in Germany that the phenomenon came to be known as the Spectre of the Brocken, after the highest mountain in this range, the Brocken. Strictly speaking, the spectre is the shadow, not the glory. You can see a glory without seeing a spectre, and vice versa. In China, where the phenomenon has been known since antiquity, devout Buddhists would make a point of visiting any mountain summit where they could witness what they termed Buddha's Aureole.

With the advent of mass air travel, everyone is almost guaranteed an opportunity to catch a glimpse of a glory, if only fleetingly. If you are flying above a layer of stratus cloud then you should be able to see a faint glory around the antisolar point, i.e. the point where the aeroplane's shadow would fall if the plane were close enough to the cloud. The brightness of the glory will vary from moment to moment depending on the thickness of the cloud at the antisolar point, and your distance from it. Sometimes, you may see a faint glory at the head of the shadow of the contrail.

If the aeroplane is close to the clouds you should also be able to see its shadow at the centre of the glory. At the same time the glory will be brighter. It may also be possible to see a fog bow beyond it. If you want to photograph it, keep in mind that the glory is much less bright than the cloud in which it is seen. You will need to over-expose the photograph.

6.5 Notes for observers of coronae and glories

1 Make an annotated sketch of the corona/glory.
2 Note down the date, time and location of the corona/glory.
3 Where were you when you saw the glory? Up a mountain, in a plane or balloon, on the ground?
4 For glory: what is the source of light? Sun, Moon, security or street lamp? For a coronae: Sun, Moon, street lamp, planet or star?
5 What is the type of cloud in which the glory is seen? What is the type of cloud or medium in which the corona is seen?
6 How many orders or rings can be seen? Count the number of red rings since these are the most easily seen.
7 Make a note of the colours that can be seen in each ring.
8 Are the rings complete? How many complete rings are there?
9 Measure the angular radius or diameter to the red edge of each set of rings.
10 Is your shadow (or that of the plane in which you are flying) visible at the centre of the glory?
11 What happens to the glory if you wave your arms, or walk about?
12 Look at the corona/glory through a polarising filter. Is the glory polarised? If so, in what direction? How does the polarising filter change the appearance of the glory?
13 Is a fog bow present beyond the glory?
14 Photographing a corona presents the same problems as photographing a halo. When the corona is around the Sun, unless you drastically reduce either the aperture or the exposure time, the brightness of the sky leads to under-exposure. When it is around the Moon, you face the opposite problem: you will have to over-expose to record the corona, during which time clouds may move.

Chapter 7 | ATMOSPHERIC HALOS

> This morning, 26th of October, being on the River coming up to London, about half an Hour past Ten, the Sun being then about twenty Degrees high, I observed a Circle about the Sun, which is by no means unusual, when the Air in Chilly Weather, such as is now, is replete with snowy Particles; which Circle was of the size it always appears in, about 23 Degrees from the Sun, and faintly ting'd with the Colours of the Iris. When this circle happens, I always look out, to see whether any other Phenomena that sometimes attend it did at that time appear, such as Parhelia, and other coloured Circles, concentrick with the Sun . . .
>
> Edmund Halley, Letter to the Royal Society, 1720,
> *Philosophical Transactions of the Royal Society*,
> vol. 3, pp. 211–12

7.1 Ice halos

Take the advice of the engaging Edmund Halley (of comet fame), and '. . . always look out . . .' for ice halos in the sky. Sooner or later you're bound to develop a nose for them, just as he did. Despite being an eminent astronomer and mathematician, and a leading member of the English scientific establishment of his day, Halley was never so busy that he hadn't time to glance up at the sky in search of halos and rainbows, and share his observations with others in his characteristically informal, easy-going manner.

Ice halos are seen in skies in which there are cirrus clouds. Unlike most clouds, these are composed of tiny ice crystals, not drops of water. Sometimes cirrus clouds are mere wind-swept tufts, and at other times they spread to fill the sky, giving it a watery appearance. There are several possible variations on the basic hexagonal shape of an ice crystal, and those in cirrus clouds are sometimes narrow columns, and sometimes thin plates. Both shapes reflect and refract light, which is why, when there are cirrus clouds in the sky, the Sun and Moon are often encircled by one or more luminous rings, sometimes accompanied by coloured spots and arcs.

The luminous rings are ice halos, and the coloured spots are known as parhelia when formed by sunlight, or parselena when formed by moonlight. There are many types of arc, and some of them are described later in the

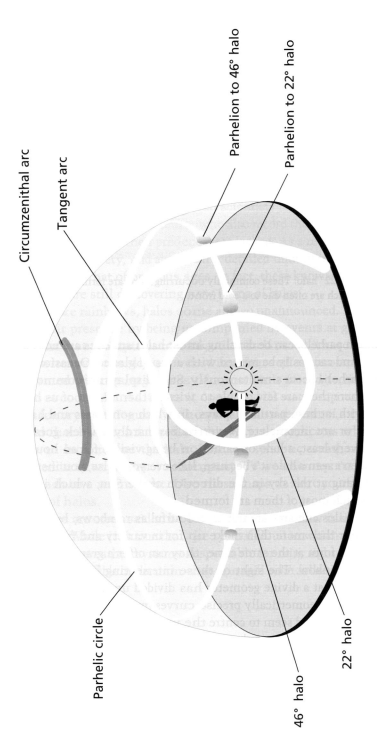

Figure 7.1 Types of ice halo. The most common forms of ice halo and their relation to an observer on the ground are shown in this diagram. The type and position of these phenomena are determined by the position of the Sun and the type of ice crystal.

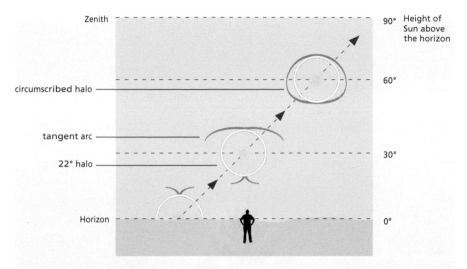

Figure 7.6 Stages in the development of a tangent arc. When the Sun is near the horizon the tangent arcs are a couple of spikes sticking out above the halo. These grow longer as the Sun rises. At the same time a couple of spikes from the bottom of the halo also grow, until, when the Sun is about 40° above the horizon, they meet to form what is known as a circumscribed halo around the 22° halo.

7.3 Upper tangent arc

Sometimes, when the Sun is close to the horizon, two arcs seem to spring from the top of the 22° halo. These are known as the upper tangent arcs and they are formed by long columnar crystals that are falling through the air with their longest axis horizontal. With increasing solar elevation, these arcs gradually increase in length, and when the Sun is about 40° above the horizon, they form a slightly oblate halo, known as the circumscribed halo, that just encircles the 22° halo. Well-developed tangent arcs are not often seen outside polar regions.

7.4 46° halo

As its name implies, the 46° halo is slightly more than twice as wide as the 22° halo. It is a very rare sight indeed, and is seldom seen complete. There are several reasons for this. In the first place, the ice crystals necessary for its formation are probably rare because they must have smooth sides and ends. The halo is formed when light is refracted between one of the sides and the base of a columnar ice crystal. The angle of minimum deviation in this situation is 46°, and it is the concentration of light at this angle that creates the ring of light that you see. However, refraction will not occur if the surfaces of the crystals are not smooth, which they won't be if the crystals have grown rapidly. Another factor that limits the occurrence of a 46° halo is that only a narrow pencil of rays from the Sun is able to pass between the two faces of the ice crystal. This light is spread over a larger area, and makes for a faint halo.

You can count yourself unusually observant, and fortunate, if you see even as much as a fragment of this large halo more than once or twice a year.

7.5 Rare halos

Halos with angular diameters other than 22° and 46° have been seen on a few occasions, and are probably caused by ice crystals with pyramidal ends. Such crystals are not particularly rare, but they seldom occur in large quantities, which is why the associated halos are seen so infrequently. They are formed by sunlight that is refracted between the pyramidal faces. The resulting halos can

Figure 7.7 Rare halos. In this photograph you can see rare ice halos of diameters of 8°, 18° and 24°. A very faint 35° halo is also visible. The 22° halo is not present. (*Photo* Pekka Parviainen)

have diameters ranging from 9° to 35° and often appear together as a series of concentric rings around the Sun. Sometimes the 22° halo may also be present.

Inevitably, there are reports of anomalous halos, on a par with anomalous rainbows. These may be genuine anomalies difficult to explain on the basis of present knowledge, but, on the other hand, they may be rare halos that have been misidentified by the observer. One possible cause for reports of anomalies is that, in the days before photography, observers were limited to verbal descriptions, backed up by sketches. These descriptions often leave much to be desired and, where unusual events are concerned, they have opened the door to endless speculation by subsequent generations of halo observers. An example of such an anomaly is known as Hevel's halo, which was included in a description of a halo display seen at Gdansk (formerly known as Danzig) on 20 February 1661 by the astronomer Johannes Hevelius. This was a particularly fine display in which most of the possible arcs and spots due to ice crystals were present. According to Hevelius, there was also an incomplete halo at 90° to the Sun. There been have no further reliable sightings of this halo, and it may be that it was really what is now known as a subhelic arc, a rare form of halo. However, the matter is still not settled one way or the other. Clearly, given the huge variety of halo types, halo observers need to be careful when they are trying to identify unfamiliar arcs in a display lest they inadvertently add to the catalogue of anomalous phenomena. One way to avoid this is to photograph the display.

7.6 Parhelia

Bright, slightly elongated streaks of light are often visible either side of the 22° halo on a line that marks its horizontal diameter. Sometimes no halo is present and only one bright patch can be seen. The streaks are due to refraction by hexagonal plate crystals. When the source of light is the Sun, they are known as parhelia or sundogs, and sometimes as 'mock suns', and when they are due to the Moon they are called parselena or 'mock moons'.

Although parhelia are often seen at the same time as a 22° halo, they are independent of it in the sense that the crystals in which parhelia are formed are not able to produce complete halos. Parhelia are thus frequently seen on their own in fibrous tufts of cirrus. Since parhelia are due to refraction, they tend to be rainbow-coloured, with their red end closest to the Sun. They can sometimes be blindingly bright, and are arguably the brightest of all optical phenomena associated with the Sun.

To create a parhelion, the refracting edges of the ice crystals have to be

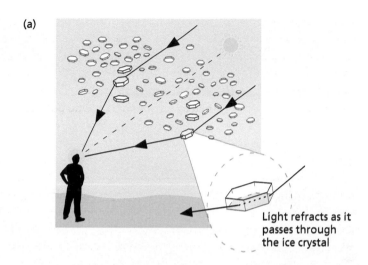

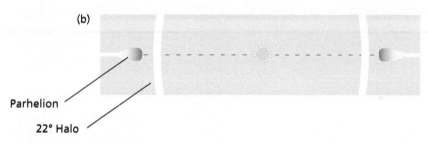

Figure 7.8 Parhelia, or sundogs, are formed by refraction through plate crystals. These fall through the air with their base more or less parallel to the ground. Diagram (a) shows the path of the rays which form the parhelion. Diagram (b) shows the relative positions of the Sun, the 22° halo and the parhelia either side of the halo.

perpendicular to the ground. This means that the crystal has to be a plate, which falls with its broad base horizontal. It follows that when we see a parhelion at the same time as a halo, the clouds in which both are formed must be made up of a mixture of crystal types. Only small columnar crystals can form halos, while parhelia are due to plate crystals.

Unless the Sun is on the horizon, sunlight cannot pass symmetrically through crystals whose refracting edges are aligned vertically. This limits the occurrence of parhelia to those times when the Sun is less than 60° above the horizon. Furthermore, the asymmetrical passage of light through the crystal means that, except at sunset or sunrise, parhelia are always more than 22° from the Sun. Hence, when they are seen together with the 22°

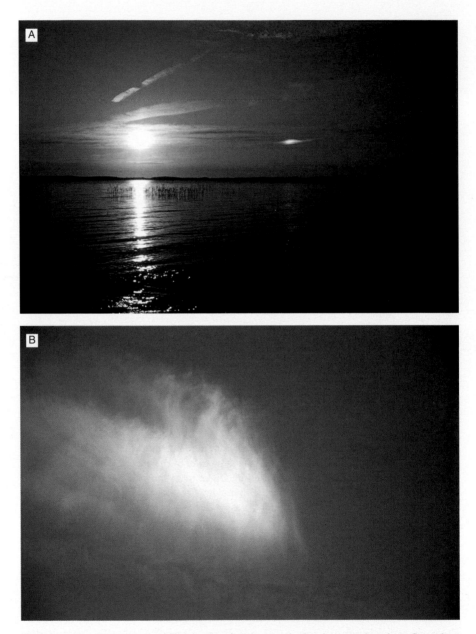

Figure 7.9 Parhelion. A parhelion, or mock Sun, may be bright enough to be reflected in water as seen in (A) (*Photo* Pekka Parviainen). In a close-up of another parhelion we can see that the sunward side of a parhelion is red (B) (*Photo* John Naylor).

halo, they will usually lie just outside it. The asymmetric passage of light also causes the sundog to smear out which is why they are elongated. Parhelia are brightest when the Sun is close to the horizon.

7.7 Circumzenithal and circumhorizontal arcs

The type of crystal responsible for parhelia can also form circumzenithal and circumhorizontal arcs. Circumzenithal arcs are much more frequent than circumhorizontal ones, though neither of them is as frequent as parhelia.

Figure 7.10 Circumzenithal arc. A circumzenithal arc is often mistaken for a rainbow, but, as you can see in this photograph, it differs from a bow in three important ways. In the first place a circum-zenithal arc is on the same side of the sky as the Sun. Secondly it curves away from the ground. Finally, the sequence of colours is opposite to that in a bow with the red edge being closer to the ground in the circumzenithal arc. (*Photo* Pekka Parviainen)

Figure 7.11 Circumhorizontal arc. This is only seen when the Sun is more than 58° above the horizon. (*Photo* John Naylor)

Circumzenithal arcs can form only when the Sun is less than 32° above the horizon. The cloud in which it is seen must be slightly more than 46° above the Sun. Thus they are seen only when the Sun is low in the sky, and there are suitable cirrus clouds near the zenith.

The relatively rapid movement of these clouds as they pass overhead means that these arcs are usually short-lived. Frequently the cloud in which a parhelion is seen will drift overhead to give a circumzenithal arc, thus giving you an opportunity to anticipate it, and prepare to photograph it.

Circumhorizontal arcs are due to light that has been refracted between a vertical face and the lower horizontal surface of an ice crystal. This is possible only for solar elevations of more than 58°. Therefore, at high latitudes circumhorizontal arcs are only seen around noon in midsummer.

Both circumhorizontal and circumzenithal arcs are the most highly coloured of all the phenomena due to refraction through ice crystals. Both rival the rainbow for colour and, indeed, are frequently mistaken for rainbows. The circumzenithal arc is slightly curved, with the edge closest to the Sun being red. The circumhorizontal arc is very slightly curved, though this may not be evident in a short segment, and forms a band of colour more or less parallel to the horizon, at least 32° above it.

Figure 7.12 Parhelic circle. Here we see a short segment of a parhelic circle and a strongly coloured parhelion. The Sun is out of the frame of the photograph on the left hand side. (*Photo* John Naylor)

7.8 The parhelic circle

Very occasionally, you may see what looks like a faint colourless band parallel to the horizon and level with the Sun. It can go right around the sky, forming a closed circle, though this is rare. This is the parhelic circle, or parhelic ring, which is created when the vertical faces of either plate crystals or columnar crystals reflect sunlight. Reflection means that there is no colour separation, which is why this arc is colourless. Plate crystals are also responsible for parhelia, so that you may initially find your eye drawn to a parhelion, only to discover that it leads on to a parhelic circle. Other halos may be present in the sky at the same time. The radius of the parhelic circle varies from 90°, when the Sun is near the horizon, to a mere 10° when it is 80° above the horizon. When the Sun is high, the parhelic circle appears as if it is a displaced 22° halo. In fact, if there is a 22° halo present at the same time, halo and parhelic circle seem like a pair of intertwined circles.

Figure 7.13 (a) A sun pillar usually appears as a faint bright steak rising into the sky above the point where the Sun has set. It occurs when there is a multitude of ice crystals in the form of hexagonal plates between the observer and the Sun. These fall slowly through the air with their largest surface more or less parallel to the ground. Each one reflects sunlight.

(b) A subsun is also a streak of light due to reflection of sunlight by plate crystals, though in this case the Sun is above the horizon and the mass of crystals in which the subsun is seen are below the observer.

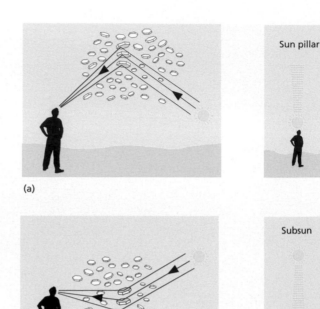

(a)

(b)

Sun pillar

Subsun

7.9 Sun pillars

Sun pillars are sometimes seen at sunset or sunrise as a faint column of light above the point of the horizon where the Sun has set, or is about to rise. The effect is due to reflection of sunlight by huge numbers of plate crystals. Since these fall towards the ground with their bases horizontal, each one acts as a microscopic mirror, reflecting the Sun's image towards the ground. In effect, you see a glitter path in the sky. Pillars of light are also formed above ground-based lights, such as street lamps, when ice crystals are formed close to the ground. This is not something you are likely to see at temperate latitudes.

7.10 Subsuns

If you face the Sun, and are able to look down on a mass of air in which there are plate crystals, you may see a bright elongated patch of light seemingly suspended in mid-air. This is a subsun. It is caused when sunlight is reflected

Figure 7.14 Left: a sun pillar appears as a faint column of light above the Sun when the Sun is near the horizon or just below it. (*Photo* Pekka Parviainen). Right: a glitter path. Sunset over the River Thames. (*Photo* John Naylor)

Figure 7.15 Light pillars. These are sometimes seen as faint streaks of light above artificial lights when the air close to the ground is full of ice crystals. (*Photo* Pekka Parviainen)

Figure 7.16 Subsun. The streak of light in the middle of this picture is a subsun. It is due to sunlight being reflected by ice crystals in the air close to the ground. Think of it as a glitter path in the air. (*Photo* Pekka Parviainen)

by the upper surface of plate crystals. Subsuns are frequently seen from aeroplanes, which they will follow for as long as the right conditions prevail, and when looking down into a valley from a mountain. Subsuns and Sun pillars are complementary phenomena.

7.11 Notes for ice halo observers

Although, in temperate climates, halos occur in clouds high above your head, in very cold climates they sometimes form at ground level. Often these ground-level halos are not noticed because they are not bright, since the depth of the cloud of ice crystals in which they are formed may not be very great. If you are somewhere where ice crystals can occur near to the ground, it's worth looking at the ground, as well as in the sky, for halos.

1. Make an annotated sketch showing the Sun, the halo (or halos), and their relationship to the horizon.
2. On the sketch you can make a note of bearings, elevation of Sun (or Moon) and angular dimensions of all halos, and other ice crystal phenomena such as parhelia. Angles can be measured using your outstretched arm (see Appendix).
3. Note down the time of day, state of sky, type of cloud and its extent.
4. What is the source of light? Sun, Moon, a street lamp?
5. Identify each type of halo. How complete is each one?
6. What colours, if any, are visible in the halos?
7. How long does each of the halos in the display last?
8. Does rain follow the appearance of the 22° halo within 12 hours?
9. Mindful of the enormous variety of halo phenomena, and to avoid adding needlessly to the catalogue of anomalies, you should photograph all halos. When photographing a halo you should:
   (a) use a slow film,
   (b) use a lens that is wide enough to accommodate it, for example a 21 mm lens for 22° halo,
   (c) take several photographs at different exposures (always underexpose to compensate for sky brightness), and spread over time (the display may change from moment to moment),
   (d) allow for the brightness of Sun's disc. This can be done in at least two ways.
      (i) Arrange for the Sun to be hidden behind some feature of the foreground. The cowl of a street light makes a good sunshield, but in an emergency use an outstretched hand, or get someone to stand so that his head just covers the Sun.
      (ii) If you want to include the Sun in the photograph, make sure that you set the exposure for the sky that does not include the Sun. When taking the picture, place the Sun in the centre of your field of view; this reduces lens flare.

A summary of ice halos

Type	Appearance	Crystal type	Comments
22° halo	Ring around Sun or Moon	Small columnar or plate crystals	The most frequently seen halo
Tangent arc	Arc at the top of the 22° halo	Large columnar crystals, basal planes vertical	Changes shape as sun rises
Circumscribed halo	Oblate halo surrounding 22° halo	Large columnar crystals, basal planes vertical	
46° halo	Circular ring around Sun or Moon	Small, perfectly formed columns	Seldom seen complete
8°, 17° and 35° halos	Concentric rings around Sun or Moon	Bullet	Very rare, faint
Parhelia to 22° halo	Bright, often highly coloured spots of light on the same level as the Sun or Moon and just outside the 22° halo	Plates	Frequently seen
Circumzenithal arc	Arc near zenith curved away from the Sun	Plates	Shows solar spectrum, red closest to Sun
Circumhorizontal arc	Horizontal arc close to horizon	Plates	Shows solar spectrum, red closest to Sun
Parhelic circle	Horizontal, white band parallel to horizon	Plates or columns	Seldom seen complete
Sun pillar	Vertical shaft of light above the Sun	Large plates	Faint, often tinged with sunset reds
Subsun	Spot of light as far below the Sun or Moon as they are above the horizon	Plates	

Chapter 8 | THE NIGHT SKY

> To persons standing alone on a hill during a clear midnight
> such as this, the roll of the world eastward is almost a palpable
> movement. The sensation may be caused by the panoramic
> glide of the stars past earthly objects, which is perceptible in a
> few minutes of stillness, or by the better outlook upon space
> that a hill affords, or by the wind, or by the solitude; but
> whatever be its origin the impression of riding along is vivid
> and abiding. . . . After such a nocturnal reconnoitre it is hard to
> get back to earth, and to believe that the consciousness of such
> majestic speeding is derived from a tiny human frame . . .
>
> Thomas Hardy, *Far from the Madding Crowd*,
> New Wessex Edition, Macmillan, 1977, p. 47

8.1 A brief history of the sky

You do not need to be told that day and night are not the same. But in what
does their difference consist? That the one is light and the other dark? Surely
there is more to them than this? There is, and it is this: the absence of
airlight allows us to see stars and planets. Their nightly drift across the sky
is much more than mere spectacle, though it certainly is that, for it is one
of the few clues we have to the fact that we inhabit a spinning sphere of solid
matter racing through a void. The immensity of this void is brought home
to us when we consider how little the sky has changed over the many thousands of years during which people have gazed up at it.

After the Sun has set tonight you will see the stars just as they appeared
3000 years ago, which is when people first began to observe the sky systematically. You will see the same stars in the same positions as they did.
It's not that the sky is changeless. No star lasts forever, and new ones are
created from time to time – though it takes as long for a star to form as the
entire time that humans have been around in one form or another: a hundred
thousand years and more. Nor are stars stationary: all stars move through
the void at huge speeds. But even the nearest stars are so far away that,
without instruments, their motion is imperceptible to us. A few thousand
years is no time at all in the life of a star, though it represents all of recorded
history. So as far as the eye can tell, the heavens haven't changed since
records first began. What have changed are our ideas about the cosmos.

The history of these ideas is long and complex, but in essence it boils down to the answer to this question: how did people come to accept that Earth is a planet in orbit about a star?

We can trace the beginnings of the answer to ancient Greece, for it is there that people first began to formulate ideas about our place in the Universe that involved bold speculation backed up by measurement and mathematics. Starting around 500 B.C., Greek astronomers gradually developed a cosmology and a mathematical model of planetary motion that reached its most developed form in A.D. 100, in the work of Claudius Ptolemy, librarian at the famous library at Alexandria.

The Greek universe was finite, eternal and, except for the Earth, unchanging. Its outermost edge was bounded by the fixed stars, and at its centre lay an immobile, spherical Earth. By our standards this universe was minute. Ptolemy estimated the distance from the Earth to the fixed stars was about 150 million kilometres, which is the same as the modern value for distance between the Earth and the Sun. The Sun, Moon and planets were said to be embedded in crystalline spheres that rotated around the Earth in the space between the Earth and the fixed stars.

The Greeks believed that all celestial motion was circular and uniform, which made it difficult for them to account for the observed motion of the planets. Ptolemy managed to overcome the straightjacket of circular motion by assuming that each planet was carried round the Earth on a series of circles. By adjusting the speed at which each of these circles rotated, he was able to approximate the observed motions of the planets. He doesn't appear to have been particularly interested in whether his system was a true description of the real situation. His principal aim was to predict the future position of planets, and in this sense his system was more of a calculating machine than an astronomical model or theory.

As outlandish and artificial as Ptolemy's universe appears to someone schooled in the astronomy of our times, his views went largely unchallenged for almost 1400 years after his death. His theory was accepted because it worked; it enabled astronomers to predict future positions of the Moon and planets relatively accurately.

The challenge was mounted, famously, by Nicholas Copernicus, a Polish astronomer. He proposed that the observed motion of planets could be accounted for more elegantly and simply by assuming that the Sun was stationary, more or less at the centre of the universe, and that it was orbited by the planets. One of these, of course, had to be the Earth. The idea that the Earth moves through the skies was something that ran counter to the common-sense physics of his time, and for which, in fact, our senses provide no direct evidence. Indeed, that the Earth moves through space was not proved conclusively until some 200 years after Copernicus died.

But Copernicus faced an insuperable difficulty when he came to work out the details of his system. The problem was that he wanted to explain planetary motion on the assumption that all celestial motion is circular and uniform, i.e. that planets move in circular paths at constant speed. In this he was merely following Ptolemy. For all his boldness in banishing Earth from its central position, Copernicus could not bring himself to defy the ancients on this score. By sticking rigidly to flawed Greek ideas about motion, he was unable to make a convincing case for the Earth's motion around the Sun, and ended up with a mechanism for planetary motion that was just as complex and arbitrary as the one he had sought to replace. Planets, as we now know, follow elliptical orbits, and any attempt to use circular motion to explain their motion is doomed to failure. At best, Copernicus was able to show that, in principle, a Sun-centred system leads to a simpler explanation of observed planetary motion than an Earth-centred one. He published his ideas in 1543, the year of his death.

In the long run, Copernicus' failure to rid astronomy of the legacy of Greek physics was of little importance because he had introduced the world to a view of the Solar System that was essentially correct. The Earth really does orbit the Sun, and it was only a matter of time before someone would stumble across the correct shape of planetary orbits. That person was Johannes Kepler, a German mathematician and astronomer. Kepler was an enthusiastic Copernican at a time when few astronomers were prepared to accept wholeheartedly a Sun-centred universe. After several years of fruitless effort trying to work out the exact shape of the orbit of Mars, Kepler finally discovered, in 1605, that the orbit of Mars, and indeed of all planets, is elliptical. But he was unable to explain *why* orbits are elliptical. This was Isaac Newton's groundbreaking achievement. Working on the assumption that a planet will travel in a straight line unless forced to do otherwise, he showed that the mutual gravitational force between a planet and the Sun could, in the right circumstances, continuously alter the path of a planet so that it travelled in an ellipse. Gravity was not discovered by Newton, but he found out how it acts between bodies. He made his theories on force and gravity known to the world in what is arguably the most important book in the history of science, *Principia Mathematica*, published in 1687.

Ideas about the Solar System and its place in the universe have, of course, changed since Newton's day. Curiously, although he supplied the scientific theories that enabled planetary orbits to be calculated, Newton himself made no useful contributions to the development of these ideas. His interest in the wider applications of his theories was primarily religious. He felt that blind chance alone could not explain how a system of planets in orbit about the Sun had come about, or how, once formed, this system remains stable over long periods of time. He believed that the fact that planets did not eventually collide with one another, or fall into the Sun, was due to divine intervention; indeed, was proof of God's existence. These ideas were a scientific dead-end. It was left to others, most notably Simon Laplace, a French astronomer and mathematician working at the end of the eighteenth century, about a hundred years after the publication of Newton's *Principia Mathematica*, to show that God was what Laplace himself called an 'unnecessary hypothesis' where the Solar System was concerned. Laplace showed that Newton's theories about gravity could account for a stable Solar System that had evolved over eons from a primitive nebula of matter to reach its present form.

Despite advances in our understanding of the nature of the Universe over the past four and a half centuries, the essential break with the past was made by Copernicus. The Copernican hypothesis, that the Earth is not at the centre of the Universe, not only provided astronomy and cosmology with a fruitful model for mapping the skies, it also provided the intellectual impetus for the development of modern physics by inadvertently posing the problem of how and why planets move around the Sun. No wonder Copernicus is celebrated as the founding father of modern science.

Four and half centuries after Copernicus, the current view is that the Solar System is but one among millions of planetary systems, which together with vast clouds of gas and dust make up a galaxy known as the Milky Way. Furthermore there are billions of galaxies within the Universe.

The ancient Greeks were wrong. The universe is not eternal, the Earth is not at its centre, and the stars do not lie at its outer limits. It is now generally agreed that space, time and energy were created in a single event of unimaginable intensity known as the Big Bang, which occurred some 12 billion years ago. Stars are created from the hydrogen and helium that arose from this event. The Universe has been expanding ever since. It is finite, though it has no edge, and will continue to expand for the foreseeable future, perhaps forever.

So, although the appearance of the sky has not changed noticeably over the past 3000 years, we no longer see it in the same way as the astronomers of the pre-Copernican era. But the fact that the Earth is not at the centre of the Universe is not something that we see or sense directly: we have to use our imagination to recognise this. Of all the sciences, astronomy is the most intriguing, and the one which most excites our imagination. The facts of astronomy, the galaxies, stars and planets, are there for us to see, and with the astronomer's help each of us can picture the Universe in our mind's eye.

8.2 Naked-eye astronomy

To our ancestors, the sky was both a calendar and a clock. For better or worse, improvements in the calendar and the invention of clocks have rendered even the most rudimentary knowledge of the stars, or the movements of the Sun and the Moon, more or less superfluous. These days, there is no practical incentive for anyone, other than an astronomer, to take an active interest in what is going on in the night sky. Consequently, most of us are not in the habit of looking at it closely as a matter of course.

Another reason why we don't look at the sky is the belief that most celestial objects and events are so remote that they can be seen only with a telescope. It's true that every single advance in astronomy since the early years of the seventeenth century, when a telescope was turned skyward for the first time by Galileo, has relied one way or another on the telescope. Without telescopes, our view of the Universe would probably be as circumscribed as that of the ancient Greeks. We would still believe that the Solar System was at the centre of the Universe, that there was only one Moon, that there were only five planets, and only a few thousand stars. Although Copernicus worked out his ideas without the benefit of telescopic observations, without the precise measurements of planetary motion for which a telescope is essential, the Copernican hypothesis might have remained just that: an unprovable theory. Telescopes have allowed astronomers to discover things that no one could have guessed at, and without which astronomy would have not have advanced much beyond Copernicus.

Compared with a telescope, the naked eye seems a puny thing. And so it is. Yet there is no better way to become acquainted with the workings of the night sky, i.e. with the movement of stars, planets and the Moon. With the naked eye you can scan the entire sky in a matter of seconds, and take in with a single glance the spectacle of a starry sky, the majesty of the Milky Way, or the thrill of a meteor shower. These are not things you can do with a telescope. Unencumbered by a telescope, you are free to roam where you will and look at whatever takes your fancy – straining to make out the Moon's surface features at one moment, noticing a halo about the Moon at the next, while delighting in the metallic and deeply shadowed landscape illuminated by the bluish light of a full Moon.

Although most of us don't look at the night sky in the hope of making discoveries, for which a telescope is essential, everyone should look at the Moon and the planets through a telescope at least once. The sight of the Moon's cratered surface, or the mottled disc of a planet, are breathtaking sights. But to become acquainted with the sky as a whole it really is necessary to look at it without a telescope. There is more than enough on view to keep anyone occupied for a lifetime. The time to think about getting a telescope is when you've exhausted all there is to see without one, or if you become particularly interested in seeing things that can't be seen with the naked eye. Even so, a telescope is not for the casual observer. Quite apart from the expense, are you going to make good use of it? Will you use it a few times and then leave it to gather dust in a cupboard? My advice is to get a good pair of binoculars instead. These are cheaper, more portable, and can be used night and day.

The important thing is not to feel that you have to approach the night sky as a budding astronomer. Naked-eye astronomy is a personal odyssey for knowledge, for spectacle and, above all, for a closer engagement with nature, as in the moment when you notice the unblinking dot of light that is a planet, and realise that there in the sky is another world accompanying Earth in an never-ending race around the Sun, and not just some abstraction in a book. An unexpected glimpse of a planet can be as exciting as an hour spent looking at it through a telescope. It all depends on your frame of mind and on what you want from the sky. Just keep in mind that you are free to look at whatever interests you, and ignore everything else.

I think that the greatest challenge we face when we look at the night sky, and the most exciting, is to free ourselves from the illusion that the Earth sits immobile at the centre of the Universe. In this sense, our experience of the night sky is no different from that of our pre-Copernican ancestors. When we look up at the stars for any length of time we become aware that the entire sky appears to rotate very slowly from east to west. This gives us the impression that it is the Earth that is at rest, and that the heavens revolve around it. Yet, at the same time, we know that this experience is an illusion. We accept, at least intellectually, that the Sun is at the centre of the Solar System, and that the Earth and planets orbit around it. Nevertheless, the geocentricism implicit in the apparent motion of the heavens is a powerful illusion, and to free oneself from it requires a considerable mental effort.

If you are to succeed in this it is not enough merely to be acquainted with the heliocentric, or Sun-centred, theory of the Solar System. This theory, at least in its essentials, is extraordinarily simple, so simple that even a child can grasp it. Yet, when confronted with a starry sky, we have no sense of the Earth's motion. Instead, the Sun and Moon appear to chase one another endlessly around the sky, and planets to weave their way among the fixed stars from one day to the next, all apparently circling the Earth. The challenge is to make the link between your first hand, Earth-centred experience, and the heliocentric reality; and this requires considerable imaginative effort. What follows is intended, in part, to help you make the necessary connections between what you see, and the reality behind it.

It has to be said that if you want to become familiar with the night sky you must be prepared to be a constant, if informal, skywatcher. Through sustained and thoughtful naked-eye observation – yes, you must think about what you see – you will gradually become familiar with the motions of celestial bodies, so that sooner or later you come to understand them for yourself, without the need to have them explained by someone else, or to refer to sky maps or data concerning planetary positions. Of course, this demands a degree of commitment. And while you don't have to go outside on every clear night, you should get into the habit of looking at the sky from time to time to keep track of what's going on 'up there'.

Every time you look at the night sky keep in mind that one of the things you are trying to understand is how the apparent Earth-centred motions of celestial bodies are linked to their real Sun-centred motions. The test of whether you have succeeded in this is whether, from the comfort of an armchair, you can picture in your mind's eye, at any time of the year, the approximate relative positions of the planets, including the Earth, and the direction in which they are moving around the Sun.

8.3 The celestial sphere

The night sky is an illusion. This is not to say that what you see is not there but rather that your interpretation of it is incorrect. All stars are so far from the Earth that it isn't possible to detect any parallax between them with the naked eye. This makes it appear as if they are all equally distant from the Earth, which in turn helps to create the illusion that we are at the centre of a huge sphere of indeterminate size.

Astronomers long ago accepted that this illusion provides the basis for a very useful fiction. For the purposes of drawing maps of the sky, and of keeping track of the apparent motions of the Sun, the Moon and planets, astronomers assume that all stars lie on the surface of a huge sphere of indeterminate size which they call the celestial sphere. The Earth's lines of latitude and longitude are projected onto this sphere so that, for example, the

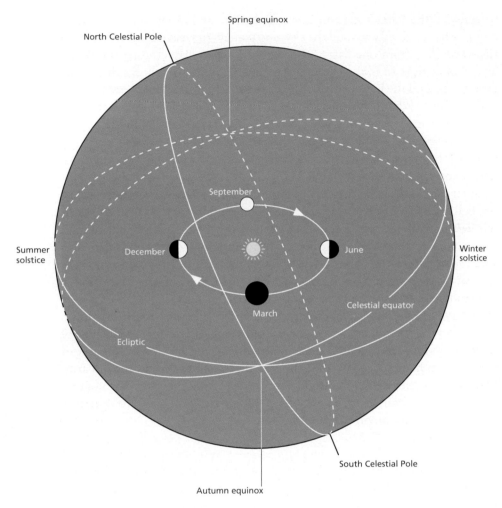

Figure 8.1 The celestial sphere is an imaginary representation of the sky centered on the Sun. The Earth's lines of latitude and longitude and its orbit about the Sun are projected onto the celestial sphere, and are used to draw star maps.

North celestial pole is directly above the Earth's geographic North Pole, and the celestial equator, which bisects the celestial sphere, lies directly above the Earth's equator. Astronomers call celestial longitude 'right ascension', or R.A., and celestial latitude they call 'declination'.

The feeling of being at the centre of a vast sphere is further reinforced by the sky's apparent motion. In the course of a single night you can see stars rise on the eastern horizon, move across the sky, and finally set at the

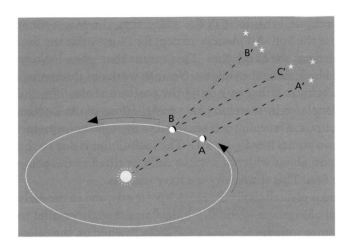

Figure 8.2 As the Earth moves from A to B along its orbit, stars that were visible due south in the midnight sky (at A') are seen to the west (at C'), while new stars are seen due south (at B'). Hence, as the Earth moves along its orbit, new stars become visible on the eastern horizon, while stars that were seen on the western horizon can no longer be seen.

western horizon. Because they keep perfect step with one another, it seems as if the sky as a whole is revolving.

In fact the celestial sphere revolves slightly faster than the rate at which the Earth spins. You will notice this if you look at a particular star, or group of stars, over a period of a week or two, and note its position relative to some landmark, such as a tree, at the same time each night. From one night to the next, the star appears to drift westward. After several months, entirely new parts of the sky become visible. This suggests that the celestial sphere rotates slightly faster than the Earth. The difference amounts to approximately four minutes per day, and is due to the fact that, as the Earth orbits the Sun, the position from which we see the night sky is continuously changing. The Earth takes 365 days to complete one revolution of 360° around the Sun. Hence, each day it moves approximately 1° along its orbit. The Earth's motion is mirrored in the Sun's apparent motion eastward along the ecliptic. Successive nights therefore bring into view a new section of the sky in the east while we lose sight of another in the west.

8.4 The ecliptic

There are two fundamental planes of reference in the sky: the ecliptic and the celestial equator. Both are consequences of the Earth's various motions. The relationship between them is the key to understanding the apparent motion of the Sun, Moon and planets.

Let's begin with the ecliptic. This is the plane of the Earth's orbit about the Sun or, to put it another way, the plane in which the Earth orbits the have not been shining for long enough to fill the universe with their light. At the same time, as the Universe expands and galaxies recede from one another, the radiation that reaches us from the more distant galaxies is shifted to longer wavelengths, such as infrared and radio waves that are not visible to the eye.

8.7 The sky beyond the equator

If you travel towards the equator from high northern or southern latitudes, the appearance of the sky changes in a number of ways. Some of the changes are expected: stars and constellations that were below the horizon in the direction of the equator come into view. This, of course, as Aristotle pointed out long ago, is a direct consequence of the Earth's sphericity. Eventually, when you arrive at the equator, some constellations, such as Orion, which we in the northern hemisphere are accustomed to see close to the horizon, are directly overhead. If you continue your journey beyond the equator you begin to notice that the whole sky has 'flipped over', and that familiar constellations and the Moon are upside-down compared with the way you are used to seeing them. Another reversal is that the winter sky of one hemisphere becomes the summer sky of the other. Orion, for example, is a winter constellation in the northern hemisphere and a summer constellation in the southern hemisphere.

There are other changes to which you have to adjust yourself when you move from one hemisphere to the other. Among these is that the apparent motion of the sky is in the opposite direction to that to which you are accustomed. Of course, the Sun, Moon and stars still rise in the east and set in the west, but you now face the opposite direction when looking at them. When you are in the northern hemisphere you are used to facing the southern sky when looking at the Sun and Moon. If you are a northerner, you come to expect, when facing that part of the sky in which the Sun and Moon are seen, that they rise on your left and set on your right. In the southern hemisphere, the Sun and Moon are seen in the northern sky, and hence rise on your right and set on your left. It can take a while to get used to the reversal.

A journey to lands beyond the equator is more than a geographical adventure. On returning to your native hemisphere, you'll find you have a more complete sense of the sky. If you are a northerner, constellations that normally peter out in the haze of the southern horizon can be completed in your mind's eye, and stars that lie far below the horizon become part of your sky, now that you've seen them and know where they lie. You are aware that Earth is surrounded by stars, something of which you may not be conscious if you've only seen the sky from one hemisphere.

Chapter 9 | THE MOON

> I was never dazzled by moonlight till now; but as it rose from
> behind the Mont Blanc du Tacul, the Mont Blanc summit just
> edged with its light, the full Moon almost blinded me; it burst
> forth into the sky like a vast star. . . . A meteor fell over the
> Dôme as the Moon rose. Now it is so intensely bright that I
> cannot see the Mont Blanc underneath it; the form is lost in its
> light.
>
> John Ruskin, Diary entry, 28 June 1844, *The Diaries of John*
> *Ruskin*, eds. Joan Evans and John Noward Whitehouse, 3 vols,
> Clarendon Press, 1956–59

9.1 Observing the Moon

The Moon is probably the only celestial object, apart from the Sun, that we
can all recognise without being prompted. It is large and bright enough to
catch our eye unexpectedly, and regularly goes though a remarkable
sequence of phases that have no parallel in nature.

There was a time when the Moon played an important part in people's
lives. Moonlight made activity possible outdoors after sunset. The first calendars were based on a lunation, the time taken for a complete cycle of lunar
phases, from one new Moon to the next. In fact, the word 'Moon' is derived
from an archaic term for measurement. Above all, it was believed that the
Moon governed all those aspects of nature that involve cycles, such as plant
life and fertility. Its waxing and waning both embodied and controlled
growth and decay. Its return to the sky at the beginning of each lunation was
seen as a resurrection, its rebirth following the death of the old Moon.
Naturally enough, people went so some lengths to arrange their lives and
activities to harness lunar powers.

Now that most of us no longer rely on moonlight, or use a lunar calendar, or believe in lunar influences, the Moon has faded into the background.
For most of us it is probably little more than a bright, mottled disc that is
occasionally visible at night; that is, if we take any notice of it at all. We are
unimpressed by the extraordinary fact that the Earth is accompanied in its
orbit about the Sun by an object large enough to be a planet in its own right.
Despite the striking changes that take place throughout the lunar cycle,
from crescent to full Moon and back to crescent, it remains a curiously

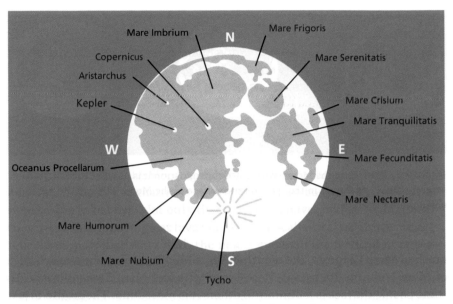

Figure 9.3 The map shows lunar features visible to the naked eye under good conditions of visibility.

they were 'upside down'. He unconsciously assumed that the North Pole must be at the top of the globe he could see, whereas he was looking at the South Pole. It was only when he realised his mistake that everything fell into place. Where the Moon is concerned, your education has probably left you to your own devices, though you will have to rise to the challenge if you wish to see the Moon as anything more than just a mottled disc. Use the map above, and binoculars, to help you find your way around the Moon.

The difficulties that we have in seeing an unfamiliar object as it really is become obvious when we are asked to describe it or draw it. A French astronomer, Camille Flammarion, famous in his day as a populariser of astronomy, conducted an experiment in 1900 in which he asked readers of a French astronomical journal to draw maps of the full Moon using the naked eye. Forty-nine people replied, some of whom were children. Perhaps the most surprising result to come out of this exercise was that not one feature was recorded by all observers. Mare Crisium, generally regarded as the most noticeable lunar feature, was recorded by only two thirds of the respondents. Flammarion's own sketch does not show it. Two sketches, made a few days apart by the same person, show the effect of libration (see section 9.16) quite distinctly, with Mare Crisium being closer to the edge of the Moon's disc in one sketch that it is in the other. In some sketches the spots appear to have been deliberately arranged to form recognisable shapes such as a human face, which was not that of the traditional man in the Moon. This suggests that, when making these sketches, some observers were influenced by what they wanted to see, rather than what really was seen, or that they wished to poke fun at Flammarion.

To achieve good results when making a naked-eye sketch of the Moon, it is essential that your distance vision is good. Before beginning to sketch, spend some time just looking at the Moon and acquainting yourself with its overall appearance. Ideally, you should make many sketches, spread over several lunations. Resist the temptation to arrange the spots into some predetermined pattern. The sketches can eventually be combined in a master sketch. Use a pencil with a soft lead and record the spots either as outlines or as shaded areas. With practice it is possible to record the subtle differences in tone between the various spots. Drawings usually improve with each successive sketch.

Everyone knows that some people have better eyesight than others. It is generally agreed that the human eye should be capable of resolving details that have an angular size of one minute of arc. This means that, under ideal conditions, someone with perfect eyesight should be able to distinguish a circle of diameter 1 cm from a square of side 1 cm at a distance of 34 metres. Few people possess eyesight as good as this and, in any case, the ability to see detail, or visual acuity, drops off with age. Fortunately, spectacles go a long way to correcting the eye's deficiencies. Visual acuity also depends on the contrast between what you are looking at and its background. When this is not great, as in the case of the lunar seas, it is difficult to see detail, even if you have good eyesight.

To subtend an angle of one minute of arc at the eye of an observer on Earth, a feature on the surface of the Moon has to be approximately 110 km across. With the naked eye you can easily make out Mare Crisium as an faint oval spot on the Moon's eastern limb because it's 430 km across, and therefore subtends an angle of approximately 4 minutes of arc here on Earth. This is equivalent to seeing an object 1 cm across from a distance of approximately 8 metres.

You can also notice, though not see, features that are below the threshold of the eye's acuity. These include the craters Tycho (diameter 90 km), Copernicus (diameter 74 km), Aristarchus (diameter 45 km) and Kepler (diameter 35 km). The reason why these craters are noticeable at all is that they are surrounded by large expanses of lighter-toned material that was ejected when they were formed, and which stands out against the darker lunar plains. The disc of the full Moon subtends an angle of 30 minutes of arc, or half a degree. This is about the same apparent size as the North American continent seen from the Moon.

Even with good eyesight, an hour or more after sunset, it can be difficult to see the spots clearly because, as the sky darkens, the Moon brightens to the point where it dazzles the eye. There are several ways in which you can prevent being dazzled by moonlight. You can look at the Moon just after sunset, when the sky is still bright. In summer, the full Moon remains close to the horizon throughout the night, and so is always much less bright than in winter, since it is seen through the lower reaches of the atmosphere. In winter, the full Moon climbs high into the sky, so look at it during twilight, when it is close to the horizon.

At night the Moon's glare can be reduced by looking at it through dark glasses or through an enlarged pinhole. Alternatively, you can reduce the size of your pupil by looking at the Moon so that its disc is seen close to a bright light: try shining a torch into your eye as you look at the Moon, or placing yourself so that the Moon is seen close to the lamp of a street light. Reducing the aperture of the eye, especially with a pinhole, has the added advantage of confining moonlight to the central part of the eye, thus avoiding distortions produced by ligaments that support the lens. Move the pinhole around until you get the clearest view.

9.4 The lunar surface

People speculated about the nature of the Moon's surface long before the advent of the telescope. There have been several schools of thought on the subject, among which were the widely held view that the spots were a reflection of the Earth's land masses, that they were shadows, or that they were due to the fact that some parts of the Moon are more transparent to sunlight than others. Nevertheless, even in antiquity, informed opinion accepted that the Moon is a world in its own right, and not a celestial mirror. Closer to our own era, both Leonardo da Vinci and William Gilbert – the only person known to have drawn a map of the Moon in pre-telescopic times – claimed that the dark areas were land masses, not reflections of the Earth's continents, and that the lighter ones were seas, like those on Earth. Galileo, who didn't believe that there were seas on the Moon, pointed out that this is actually the reverse of what would be seen if there were seas on the Moon: from above, a body of water usually looks darker than land.

Galileo was the first person to look at the Moon through a telescope, and identify some of its major features more or less correctly. His observations were made during the winter of 1609/10, using a crude telescope that he had constructed himself, and which was capable of a magnification of 30 times. With this instrument he was able to make out the huge plains of lava that are visible to the naked eye – which later came to be called 'maria' (singular: 'mare', a word that means 'sea') – mountains, and many of the larger craters that cover much of the south-eastern quarter of the Moon. In *The Sidereal Messenger*, his own account of these and other telescopic discoveries, Galileo wrote

> . . . consequently anyone may know with the certainty that is due to the use of our senses, that the Moon certainly does not possess a smooth and polished surface, but one rough and uneven, and just like the surface of the Earth itself, is everywhere full of vast protuberances, deep chasms and sinuosities.

Galileo Galilei

Galileo claimed that many of the features that he saw through his telescope were similar to those found on Earth, though he didn't know what to make of craters, describing them as small spots, or cavities. The first person to make a stab at working out what these 'cavities' are was Robert Hooke, who came to the conclusion that they were due to volcanic eruptions, a view that prevailed among most astronomers until the 1960s. The current view is that almost all lunar craters are due to objects that collided with the Moon in the distant past. Hooke was aware that craters could be formed by impacts. He carried out an experiment in which he dropped objects into wet clay, but despite the similarity between the appearance of the craters produced in this way and those he could see on the Moon, he could not accept that large objects had collided with the Moon.

The other side of the Moon, the so-called 'far side', remained unseen until January 1959 when it was photographed by Luna 3, a Russian probe. There are hardly any maria on the far side, and its overall appearance is very different from that of the 'near side' with which we are familiar. Had the far side been the one that always faces Earth, the Moon would have appeared almost featureless to the naked eye.

Galileo was a competent draughtsman, and made several telescopic sketches of the Moon showing some surface details. But he never drew the full Moon, and made no attempt to give names to its features. The first map of the Moon that deserves the name, one that clearly showed and named the plains, mountains and craters, was drawn by Michael van Langren, and published in 1646. However, the names he used have not survived, and modern

Moon maps use a system of naming devised by Giovanni Riccioli, a Jesuit astronomer. In this system, craters are named after scholars: those in the Moon's northern hemisphere were named after the ancients (e.g. the crater Plato), and those in the southern hemisphere after Riccioli's contemporaries (e.g. the crater Grimaldi – after Franceso Grimaldi, Riccioli's pupil and also a Jesuit, who drew the map used by Riccioli). Mountain ranges were named after those on Earth (e.g. the Apennines). The maria were named after states of the weather (e.g. Mare Imbrium: the Sea of Rains). Of course, since Riccioli's day, new features have been discovered and named (though, by and large, his system has been adhered to). Even on modern maps, however, only major features have names.

In a sense, the ancients who believed that the Moon held up a mirror to the Earth were not so wide of the mark – but it is a reflection of the Earth as it might have been over 4 billion years ago. Planetologists are now generally agreed that the Moon was created when the planet-sized object that eventually became Earth was struck a glancing blow by another huge planet the size of Mars some 4.5 billion years ago. During this collision, material from the outer layers of both objects was hurled into space, where some of it rapidly came together to form the Moon. The remains of the colliding body coalesced with the planet it had struck, and together they formed the Earth. There followed a period of several hundreds of millions of years during which both the Earth and the Moon were bombarded by vast numbers of bodies, large and small, which were then plentiful in the Solar System. These produced huge numbers of craters on the surface of both the Earth and the Moon. On the Moon the largest impacts may have triggered vast lava flows that gave rise to the lunar maria. This period of intense bombardment ended some 3.5 billion years ago. Since then, the Moon's surface has been continuously bombarded by tiny meteoroids, which have gradually eroded its surface. There have also been impacts from larger bodies: the craters Copernicus and Tycho were formed 800 million and 100 million years ago respectively. Nevertheless, almost all the major features on the Moon's surface that we see today are essentially the same as they were 3.5 billion years ago. Here on Earth, tectonic activity and erosion due to wind and water have completely transformed the Earth's surface and erased any traces of all but the most recent craters. The Earth's surface is substantially different now from the way it was even 100 million years ago, let alone 3.5 billion years ago. Because the Moon's surface today is a reminder of an earlier era when both it and the Earth underwent massive bombardment by smaller bodies, in that sense it holds a mirror to the Earth's distant geological past.

9.5 The best time to look at the Moon

The Moon is an endlessly fascinating object, and provides interesting views whatever its phase, depending on what you want to see. If it is earthlight (see section 9.8), then you should look at the Moon when it is a crescent, either waxing or waning. If you want to look for the man in the Moon, and the other figures in the spots, you should look around full Moon. And if you want to catch sight of the Moon in daylight, there are times of the year when you will be wasting your time.

If you want to look at the Moon's surface using binoculars or a telescope then, arguably, some of the most dramatic views are to be had when it is a crescent either at the start or at the end of a lunation. As it waxes, it grows brighter, and the contrast between adjacent features is diminished to the point where, around the time of the full Moon, it is difficult to make out most craters. What makes a crescent Moon an interesting sight are the elongated shadows cast by mountains and craters close to the terminator. These shadows emphasise the roughness of the lunar surface and make craters and mountains much more evident than they are at other times of the lunation. The difference that these deep shadows make to the visibility of lunar features is very obvious if you look at the Moon through a telescope on successive days. Craters that stand out clearly one day appear to vanish the next. At the same time, the illuminated peaks of the tallest mountains and the rims of some craters that lie just beyond the visible portion of the Moon's surface give the terminator a ragged edge.

The main thing to bear in mind when choosing a good time to look for these things is that you stand a better chance of seeing them when the Moon is high in the sky. And this depends on its apparent path across the sky relative to the horizon, which changes throughout the year. You can find out more about the Moon's motion in section 9.18.

In the northern hemisphere, the very best time to observe the waxing crescent Moon is in the evening around the time of the spring equinox. At this time of the year the crescent is always high above the horizon long after the Sun has set. The waning crescent is similarly well placed shortly before sunrise around the time of the autumn equinox. In the southern hemisphere the situation is reversed and the waxing crescent Moon will be at its highest above the horizon on autumn evenings.

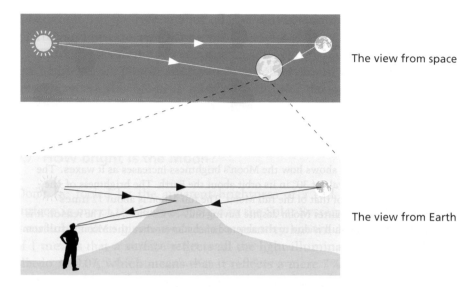

The view from space

The view from Earth

Figure 9.7 Estimating the brightness of the Moon's surface. The brightness of the full Moon can be estimated by comparing the full Moon with some large feature on Earth when the Moon is close to the horizon. Since the Moon is 400 times closer to the Earth than it is to the Sun, sunlight that illuminates the Earth has travelled more or less the same distance as that which illuminates the Moon. Furthermore, direct sunlight is attenuated to the same degree in its passage through the atmosphere as is sunlight reflected by the Moon. Hence sunlight illuminating the Earth's surface has the same intensity as that which we receive from the Moon. Estimating the Moon's brightness is thus reduced to a matter of finding a surface here on Earth that has the same brightness as the Moon.

> the Moon setting behind the grey perpendicular façade of the Table Mountain illuminated by the Sun just risen in the opposite quarter of the horizon, when it has scarcely been distinguishable in brightness from the rock in contact with it. The Sun and Moon being at equal altitudes and the atmosphere perfectly free from cloud or vapour, its effect is alike on both luminaries.

J. Herschel

The reasoning behind Herschel's comparison is as follows. The distance between the Moon and the Earth is approximately 400 times less than that between the Earth and the Sun. Hence sunlight illuminating the Moon travels almost the same distance as that which illuminates objects on the Earth. Herschel was therefore justified in assuming that both bodies are illuminated by sunlight of the same intensity. Furthermore, under the condi-

Figure 9.8 Full Moon. When a full Moon rises in the northern hemisphere its axis is almost parallel to the horizon as shown here. (*Photo* Pekka Parviainen)

tions in which he made his observation, namely with both Moon and Sun close to the horizon, sunlight which illuminates objects on the Earth's surface is attenuated by its passage though the atmosphere to the same degree as is the light that reaches the eye from the Moon. Hence, if the Moon appears to have the same brightness as the object with which it is being compared, it must be because both have a similar albedo. Note that the comparison should be made at full Moon, and that the Moon's disc should be visible alongside the bit of the Earth's surface with which it is being compared. The object itself must be directly illuminated by the setting Sun, and both eastern and western horizons should be equally clear of haze or clouds.

Figure 9.9 When the full Moon is close to the horizon it may be reddened and distorted just like the Sun. (*Photo* Pekka Parviainen)

In addition to the objective changes outlined above, there is a subjective dimension to the Moon's brightness. You will have noticed that the Moon undergoes a startling increase in brightness shortly after sunset. The increase is particularly striking around full Moon. An hour or two before sunset, the Moon seems little brighter than the sky. But as the sky darkens, the Moon loses its pallor, and shines ever more brightly in the gathering gloom. Yet the amount of sunlight being reflected by the Moon remains constant. What changes is the brightness of the sky: as it grows darker, the Moon appears to grow brighter.

In fact, any object will appear to grow brighter when its surroundings are darkened if the object continues to shine with undiminished intensity. Stars and planets – with the occasional exception of Venus and Jupiter (see section 11.6) – are not normally visible to the naked eye during daylight because the daytime sky is brighter than any star or planet. After sunset, as the sky's brightness diminishes, stars and planets make an appearance, one by one, beginning with the brightest. To the casual observer, it appears as if these bodies grow brighter as twilight gives way to night, though, of course, they don't; it is the sky that gets darker. At sunrise, the situation is reversed: as the sky gets brighter, stars and planets fade from view – though, of course, they don't cease to shine. Artificial lights such as torches, streetlights and car headlamps also appear to be much brighter after dark than they do while it is still light.

The apparent change in brightness in all these cases is due to the fact that the eye does not respond to brightness in absolute terms, but does so by comparing the brightness of an object with that of its surroundings. Of course, at high latitudes, the night sky is not as dark in summer as it is in winter and so a full Moon seen on a June or July night looks distinctly mellow when compared with the harsh silvery light of a December or January full Moon. In summer a full Moon is even less bright because it is never far from the horizon and so is seen through a considerable depth of atmosphere. Atmospheric scattering reduces the proportion of shorter wavelengths in the Moon's light, making it appear yellower and less bright.

9.8 Earthshine

For several days before and after new Moon, when the Moon is a slender crescent, either before sunrise or after sunset, if the sky is relatively dark, you can often make out the outline of the segment of the Moon's disc that is not directly illuminated by the Sun. With binoculars it is even possible to make out the dark spots of the maria.

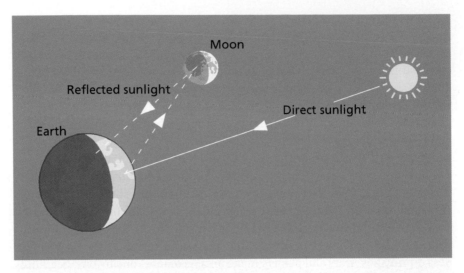

Figure 9.10 Earthshine (or earthlight) is the name given to sunlight reflected by the Earth at the dark side of the Moon and reflected back again by the Moon to Earth. It is most noticeable during the first and last few days of a lunation, when the Moon appears as a crescent.

The Moon is not intrinsically luminous, and so is not the source of the light that makes its night side visible to us. The origin of the light that illuminates this portion of the Moon is sunlight reflected by the Earth. The Earth's albedo is about six times greater than that of the Moon, and, of course, the Earth has a much greater surface area than the Moon. Furthermore, seen from the Moon the Earth also goes through a cycle of phases, just like those that we see the Moon go through. The Earth's phases, however, are the opposite of the Moon's phases. In other words, at new Moon, the Earth is full, and vice versa. Hence, when the Moon is a crescent the Earth is gibbous, and enough sunlight is reflected by the Earth to make the night side of the Moon visible.

This phenomenon is known as 'earthshine' or 'earthlight', and sometimes as the 'old Moon in the new Moon's arms', or the 'Lumen Cinereum'. It is most noticeable at the beginning and at the end of a lunation because, as the Moon waxes and grows brighter, the Earth wanes and grows dimmer. In the northern hemisphere, earthshine is seen most clearly when a crescent Moon is high above the horizon after sunset around the time of the spring equinox, or before sunrise around the time of the autumn equinox. It becomes more noticeable as the sky darkens.

It has long been known that earthshine is invariably brighter, and thus

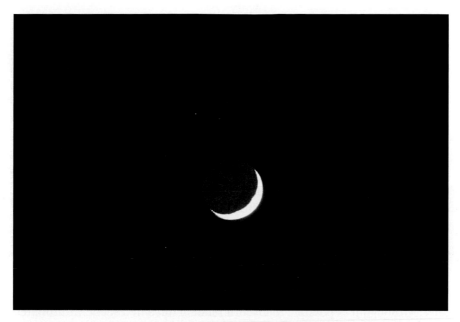

Figure 9.11 Earthshine. Sometimes when the Moon is a crescent its unilluminated portion is visible because of light reflected by the Earth. (*Photo* John Naylor)

more noticeable, when the Moon is waning than when it is waxing. One reason for this is that maria make up a smaller proportion of the eastern hemisphere of the Moon than they do of the western hemisphere. The eastern hemisphere lies within the night side of the waning Moon. Hence the night side of a waxing Moon, being darker, reflects less of the Earth's light than the night side of a waning Moon.

Earthshine is not the only light that reaches the Moon from the Earth. During a total eclipse of the Moon, the Moon takes on a reddish hue from light that has passed through the Earth's atmosphere. This light has been reddened due to scattering within the atmosphere. You can find out more about the colour of the Moon when it is eclipsed on page 238.

9.9 Moonlight

The amount of light reflected by the Moon in the Earth's direction varies with its phase. Moonlight becomes noticeable only at night, when the sky is dark, which is why we are aware of it only between first and last quarter, when the Moon spends a significant amount of time in the night sky. Before widespread public lighting, moonlight was an important source of light, making activity possible at night. But moonlight makes it difficult to see faint stars, even with a telescope, so astronomers don't like moonlight. It interferes with their work, so they arrange their schedules to take account of the Moon's phases.

The belief that the Moon affects human behaviour adversely is very ancient. Moonlight, particularly, was said to have a malign influence. It was widely believed that if it shone on your face while you were asleep, it would weaken your eyesight, or even drive you mad. The original meaning of 'lunatic' is someone who has been driven mad by the Moon. Lunacy was supposed to be most likely at full Moon. But there isn't much, if any, hard evidence that the Moon has any effect at all on people, good or bad.

The light of the full Moon is approximately half a million times less bright than that of the Sun. A surprising consequence of this is that a sheet of white paper illuminated by moonlight is much less bright than a sheet of black paper illuminated by sunlight. At such low levels of illumination, the cone cells in the eye, which enable us to perceive colour, do not operate reliably. Although it is possible to discern reddish hues in a scene illuminated by the light of a full Moon, particularly if the coloured surface occupies a large part of your visual field, under such conditions, sight is mainly by rod vision (see section 4.10). A moonlit scene thus appears devoid of colour. Nevertheless, since moonlight is reflected sunlight, it has the same spectrum of colours as sunlight, and if the exposure is sufficiently long, say, half an hour or more, a colour photograph of a moonlit scene will reveal a landscape that is all but indistinguishable from its daylight counterpart. In such a photograph, grass is green and the sky is blue, with only tell-tale star-trails to give the game away.

Many people find that moonlight has a greenish or bluish cast. This is particularly evident around full Moon, when the Moon is at its brightest. It has been suggested that this effect is due to the Purkinje effect (see section 4.10). But this can't be correct since the Purkinje effect is the result of switching from cone, or colour vision, to rod, or brightness vision. Rods don't respond to colour, so rod vision is monochromatic. I have not found alternative explanations of why moonlight appears blue, and so I can't say whether the effect is physical or physiological.

You can test these things for yourself. Go outdoors or, better, take a walk on a night when the Moon is almost full and high in the sky. Does everything appear to be bathed in a bluish light? Notice how profound shadows are (you can find out more about shadows in moonlight in section 2.1). Is it possible to make out colours? You could use colour chips from a paint catalogue to answer this. Don't look at the chips before you start. When in moonlight, write down the colour beside each chip. Does the apparent size of the chip affect the outcome? How well can you see by moonlight? It is often claimed that one can read a newspaper by the light of a full Moon. This, of course, is easy to verify. What about objects? Look at familiar objects and compare the amount of detail seen in sunlight with that seen in moonlight.

9.10 The Moon illusion

You have probably noticed that the Moon, particularly around full Moon, appears larger when it is at, or near, the horizon than when it is high in the sky. Some people also report that this enlarged Moon appears to be unexpectedly close. Despite appearances, however, you are actually slightly further from the Moon when you see it on the horizon than when you see it overhead. Its climb from horizon to zenith is due to the Earth's rotation through a quarter revolution, which brings you closer to the Moon by a distance equal to the Earth's radius. The tiny change that this brings about in the Moon's apparent diameter is not perceptible to the naked eye, and so any changes in either apparent size or distance as the Moon climbs into the sky are illusory. An apparent enlargement at the horizon is also noticeable with both the Sun and the constellations.

The phenomenon is known as the 'Moon illusion', and it has been the subject of speculation since antiquity. Aristotle believed that it had a physical cause, and suggested that it was due to looming, i.e. to what we would call abnormal refraction by the atmosphere. This explanation was rejected several hundred years ago when the phenomenon was recognised as being an optical illusion, and hence a problem of perception. Nevertheless, it has proved to be a very complex illusion and, even today, there is no generally accepted explanation for it. Indeed, there is even disagreement about the nature of the illusion. For example, it is often assumed that it is the enlarged horizon Moon that is illusory. But it has been suggested that the horizon Moon is the one seen normally and that it is the zenith Moon that looks smaller than it should. Most explanations, however, begin by assuming that the illusion occurs at the horizon. A widely accepted explanation for horizon enlargement is based on the fact that the horizon appears to be further away from us than the zenith does. Experience teaches us that a distant object that appears to have the same size as one that is near at hand must be the larger of the two. The mind thus enlarges the 'horizon' Moon in accordance with this unconscious assumption.

Figure 9.12 Moon illusion. The disc of the Moon appears larger when seen close to the horizon than it does when it is high in the sky. There is no physical reason for this so this really is an illusion.

The illusion remains unexplained because changes in the Moon's apparent size and distance are probably a composite of several illusions and, in any case, we do not have a good enough understanding of visual perception to deal adequately with any of these. In other words, at the time of writing, there is no generally agreed explanation for this illusion.

9.11 A Blue Moon

A Blue Moon is supposed to occur so rarely that it has become a byword for an unusual event. In fact, the term 'Blue Moon' has two meanings. In the last fifty years, among astronomers, a Blue Moon has come to mean the second full Moon that occurs in the same month as a previous one. This is quite a rare event and takes place about once every three years. The rarity of this event, however, is simply the result of our Sun-based calendar, which is not designed to keep in step with the Moon's phases.

The second, altogether more interesting, use of the term refers to a change in the Moon's colour. Under certain, rare, atmospheric conditions both the Sun and the Moon take on a faint bluish hue. This is due to a particular form of selective scattering by particles in the atmosphere and not to something happening on the Moon. The atmosphere itself is composed of molecules of gas. These molecules are smaller than the range of wavelengths of visible light, and they scatter shorter wavelengths to a greater degree than longer wavelengths. Such scattering is sometimes called Rayleigh scattering (see section 1.2), and is the cause of the blue colour of the sky. Rayleigh scattering is responsible for making both the Sun and Moon appear orange when they are close to the horizon (see sections 1.3 and 1.4 for more on Rayleigh scattering). If the scattering particles are much larger than the wavelength of light, they scatter all wavelengths equally. This is the case with the drops of water that make up clouds, which, therefore, are white. If, however, the size of the scattering particles is more-or-less the same as the wavelength of light, longer wavelengths are scattered to a greater degree than shorter ones and the transmitted beam rather than the scattered beam takes on a bluish hue.

Particles of the necessary size, and in sufficient concentrations to cause this type of scattering, are rare in nature. Nevertheless they can and do occur. In 1950, for example, smoke from massive forest fires in Canada was carried across the Atlantic and imparted a blue hue to the Moon seen in Europe. The cause was tiny drops of oil distilled from the resin of the burning trees. Blue Moons and Suns were also seen in the months following the eruption of Krakatoa.

Blue Suns are sometimes seen during dust storms. For example, it is reported that the Sun sometimes turns blue in northern China when the dust in such a storm is the fine loess from Mongolia. The longer wavelengths that are scattered by particles of loess impart a reddish hue to the landscape. Blue Suns have also been seen during dust storms elsewhere.

> Visiting Fathepur in Rajastan, on 28 May 1989 to supervise crop trials, I was caught in a sandstorm which raged all morning and afternoon, obscuring the Sun completely. When it appeared briefly at about 16.30, I was a little disconcerted to find that the disc was bright blue but after 10 to 15 minutes it was again obscured by dust and did not reappear that day . . .
>
> J.M. Peacock

This phenomenon is not confined to Earth: planetologists have realised that the rosy hue of the Martian sky photographed by probes on the planet's surface is due to the presence of fine dust in its atmosphere. There is a photograph of a Martian sunset in which the Sun is noticeably blue.

9.12 Sidereal and synodic months

The Moon takes 27.32 days to make a complete orbit around the Earth. This period is known as the sidereal month. 'Sidereal' means 'relative to the

Figure 9.13 Synodic and sidereal months. The length of a lunation is greater than the time it takes the Moon to complete an orbit of the Earth because the Moon must complete more than one orbit to bring it back in line with the Sun, i.e. the time taken from one new Moon (at A) to the next (at C) is a few days more than the time taken to complete an orbit. The Moon has completed one orbit when it has returned to the same position in the sky (at B) where it is once again seen against the same group of stars.

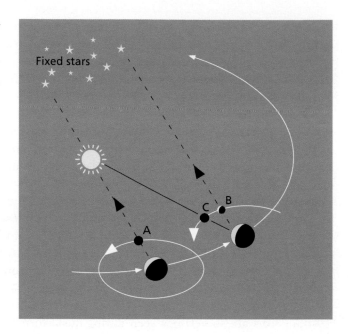

stars', and comes from Latin. However, an Earth-bound observer must wait between 29 and 30 days to see the Moon go through a complete cycle of phases because these are determined by the relative positions of the Sun, Moon and Earth. This longer period of time is the synodic month and it is the basis of lunar calendars. 'Synodic' means 'meeting' or 'conjunction' and comes from Greek. The synodic month is the time taken for a lunation. The Moon goes through a complete cycle of phases in the course a lunation.

The synodic month is longer than the sidereal month because the Earth and Moon are not stationary, they orbit the Sun together. The effect of this is shown in figure 9.13. The Moon's phases are determined by the relative positions of the Earth, the Moon and the Sun. New Moon occurs at the moment when the Moon lies directly between the Earth and the Sun. In order to return to the same relative position, the Moon must complete slightly more than one revolution around the Earth because the Earth itself moves around the Sun.

9.13 Finding the Moon in the sky

It's very satisfying to be able to work out where you have to look to see the Moon on a particular day. But, to a casual observer, it can easily seem as if there isn't much rhyme or reason to when and where the Moon appears in the sky. Partly this is because, if the sky is overcast for several days in a row, we may miss seeing it altogether, and won't be able to see where each phase appears in the sky in relation to the others. In fact, the Moon moves across the sky like the Sun, rising in the east and setting in the west.

To make sense of the Moon's apparent motion you need to know a couple of things. The Moon orbits the Earth in a plane that is inclined at 5° to the ecliptic. Hence its apparent path across the celestial sphere, broadly speaking, is similar to that of the Sun. The Moon, however, takes just over 27 days to go around the celestial sphere once, whereas the Sun takes a whole year to do the same thing.

Since the Moon goes all the way around the celestial sphere in the course of a lunation, the points on the horizon at which it rises and sets during this period will swing between more or less the same extremes as those of the Sun do in the course of a year (see section 8.5 for the Sun's yearly motion). At the start of a lunation, the Moon lies directly between the Earth and the Sun. A young Moon therefore sets at almost the same point on the horizon as the Sun. But half a lunation later, the full Moon rises at a point on the horizon close to that at which the Sun will rise six months hence. Thus, in summer the daily path of the full Moon across the sky is similar to that taken by the Sun in winter: for observers in the northern hemisphere, it rises far to the south and remains close to the horizon as it crosses the sky from east to west. If you use the fact that the apparent motion of the Moon is linked to the Sun's position on the ecliptic, you won't find it too difficult to make a rough prediction of the path of the each of the Moon's key phases across the sky during each of the four seasons (see figure 9.14). And, of course, we always have an approximate idea of where the Sun is on the ecliptic since this determines the season. For example, in midsummer, the Sun is as far north of the celestial equator as it can get, and in autumn it crosses the celestial equator from north to south.

As an example of how you can work out the Moon's daily path across the sky in advance, consider a lunation that begins when the Sun is at, or very close to, the autumn equinox. Every day after the start of the lunation, the Moon moves eastward away from the Sun, more or less along the ecliptic. This half of the ecliptic lies south of the celestial equator, and so during the first half of this lunation, the waxing autumn Moon hugs the southern horizon. This makes it a difficult object to see, since it may be hidden from view behind buildings, trees or hills. During the second half of the lunation, however, the Moon moves into the northern celestial hemisphere, and so the waning autumn Moon is seen high in the sky. The situation is reversed

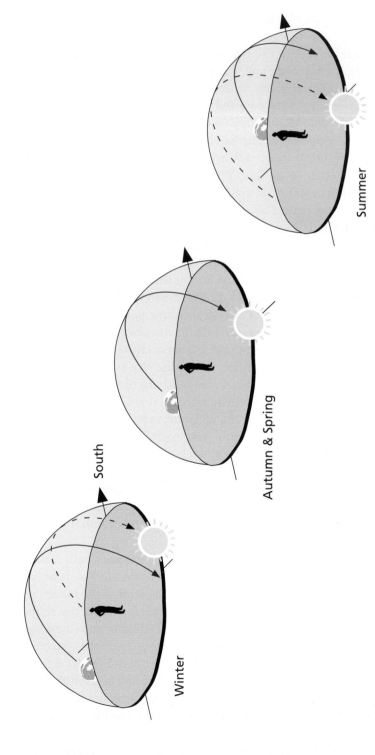

South

Winter

Autumn & Spring

Summer

Figure 9.14 The apparent path of the full Moon across the sky at different seasons. During autumn and spring, the full Moon follows more or less the same path across the sky as the Sun. In winter, the Sun remains close to the horizon throughout the day, while the Moon rises high into the sky at night. During summer, the situation is reversed, and the full Moon is never far from the horizon, while the Sun climbs high into the sky.

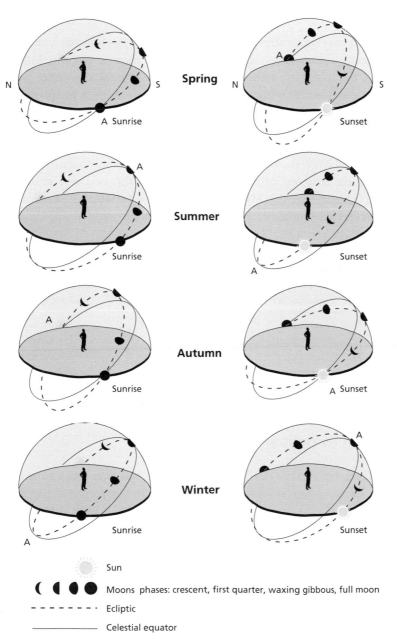

Spring

Summer

Autumn

Winter

Sunrise

Sunset

Sun

Moons phases: crescent, first quarter, waxing gibbous, full moon

Ecliptic

Celestial equator

Figure 9.15 These diagrams show the approximate position of the Moon's phases during the first half of a lunation at sunrise and at sunset. 'A' marks the spring equinox. Notice that the sky at sunrise during spring is the same as the sky at sunset in autumn, and vice versa. The same is true for summer and autumn.

The Moon's daily path across the sky is always parallel to the celestial equator, whatever its phase.

for the lunation that starts when the Sun is at the spring equinox: a waxing Moon in spring always rises higher in the sky than a waning Moon. Observers in the southern hemisphere see these events the other way around: a waxing crescent is higher in the sky in autumn than in spring.

The season in which each key phase reaches its maximum or minimum altitude above the horizon, i.e. when it culminates, is given in the table below. In this table 'maximum' means that the Moon will culminate between 18.5° and 28.5° above the celestial equator, and 'minimum' means it will culminate between 18.5° and 28.5° below the celestial equator. 'Intermediate' means that the Moon will culminate between 5° above and 5° below the celestial equator. 5° is, of course, the inclination of the Moon's orbital plane to the ecliptic.

Altitude of key phases at culmination in different seasons

	First quarter	Full Moon	Last quarter
Spring	Maximum	Intermediate	Minimum
Summer	Intermediate	Minimum	Intermediate
Autumn	Minimum	Intermediate	Maximum
Winter	Intermediate	Maximum	Intermediate

9.14 Moonrise and moonset

A lunation lasts, on average, 29.5 days and so the Moon appears to move across the celestial sphere at an average rate of approximately 13° per day. In other words, on each successive day, the Moon will rise with those stars that were 13° to the east of it at the previous rising. At the same time, since the Earth itself orbits the Sun, the Sun also appears to move eastwards across the celestial sphere, though at the much slower rate of 1° per day. Thus a waxing Moon moves away from the Sun at an average rate of approximately 12° per day, and moonrise should be delayed by approximately 50 minutes from one day to the next (24 hours/29.5 days = 50 minutes per day). As a rough guide, the first quarter Moon rises some 6 hours after sunrise, the full Moon rises some 12 hours after the Sun, and the last quarter Moon rises some 18 hours after the Sun. See the table on page 214 for more on the visibility of lunar phases.

Moonrise, of course, is when the Moon rises above the eastern horizon. Outside the tropics, the interval between the times of moonrise on successive days vary widely throughout a lunation. This is because the Moon, whatever its phase, rises fastest when it reaches the point along its path around the celestial sphere where it crosses the celestial equator from the southern hemisphere to the northern hemisphere, i.e. when it is at the point on the ecliptic known as the spring equinox (see figure 8.1). This point lies in the constellation of Pisces. When the Moon is close to this point, the delay between successive moonrises is small (some 15 minutes at the latitude of London, and less at higher latitudes). A couple of weeks later in the same lunation the Moon reaches the autumn equinox, which is in the constellation of Virgo, and passes from the northern hemisphere to the southern one. The delay between successive moonrises then becomes greatest (about 90 minutes at the latitude of London and more at higher latitudes). At moonset, the relationship is reversed: a Moon that rises quickly, sets slowly and vice versa.

The reason why the interval between successive moonrises varies throughout a lunation is that the angle between the ecliptic and horizon changes continuously. To understand how the angle between these affects the rate at which the Moon rises, consider the well-known phenomenon of the harvest Moon.

Harvest Moon is the name given to a full Moon that occurs when the Sun is at the autumn equinox. If a full Moon is to occur when the Sun is at this point on the ecliptic, the Moon must be at the opposite side of the ecliptic, i.e. at the spring equinox in the constellation of Pisces. We saw above that, when the Moon is near to the point on the celestial sphere known as the spring equinox, the interval between sunset and moonrise on consecutive days is small. If the Moon happens to be full, or almost full, when it is in Pisces, moonlight quickly replaces that of the setting Sun for several days in succession. Daylight thus gives way to moonlight on successive evenings for up to a week, allowing farmers – so it is claimed – to work long after sunset during harvest time (hence the name: harvest Moon).

To understand why the Moon rises rapidly at this time of year, look at figure 9.16 as you read though the explanation given below.

To show the difference that the time of year makes to the interval between successive moonrises, the diagram shows the track of the Moon along the ecliptic both in autumn and in spring. Each day the Moon moves along the ecliptic about 13° in an easterly direction. Let's assume that full Moon (labelled A in the diagram) occurs at the moment when the Moon crosses the celestial equator. If this happens at the autumn equinox, the Moon's position at the same time on the next two days will be those labelled

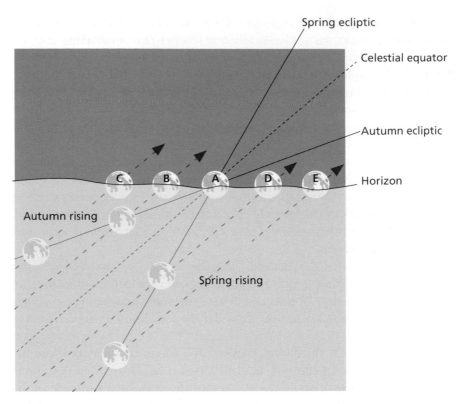

Figure 9.16 The harvest Moon. The Moon always rises along a path that is parallel to the celestial equator, i.e. along the broken lines in the diagram. It takes less time for the Moon to rise to the horizon when it has reached the same point in its orbit in autumn than in spring, which is why the difference in the time of moonrise is less around the autumn equinox and greatest around spring equinox.

B and C in the diagram. On the other hand, if full Moon occurs at the spring equinox, the subsequent daily positions of the Moon will be those labelled D and E.

Moonrise, however, is not due to the Moon's motion; it is brought about by the Earth's rotation. And since this takes place about an axis that is perpendicular to the equator, the path taken by the rising Moon is also parallel to the celestial equator. This path is shown in the diagram by a solid line. It can be clearly seen that on successive autumn evenings the Moon doesn't have as far to travel to reach the horizon as it does on successive spring evenings. The delay in moonrise from one day to the next is therefore far less in autumn than it is in spring. Consequently in spring the twilight glow has already faded before moonlight can take its place, whereas in autumn, moonlight can augment twilight for several evenings in succession.

Like all astronomical phenomena that depend upon the motion of the bodies involved, you will probably find that you have to turn this one over in your head for a while before you really understand it. You may find that a planisphere or a computer simulation of the night sky is a great help in visualising the motion of a rising and a setting Moon.

9.15 The lunar day

As we have already noted, we always see the same pattern of spots on the Moon's surface, because we always see the same side of the Moon. This side is known as the 'near side'. The opposite side, the one that is never seen from Earth, is known as the 'far side'.

The reason why we only ever see the near side is that the Moon rotates once on its axis in the same time as it takes to complete its orbit around the Earth, i.e. the lunar day lasts one sidereal month. If these were not the same, if the Moon rotated more slowly or more quickly than it actually does, then, given time, we would be able to see the entire lunar surface though, depending on the ratio of the lunar day to the sidereal month, this might take more or less than a single lunation.

The Moon rotates about its axis in an anti-clockwise direction seen from the North celestial pole, i.e. from east to west relative to the Earth's surface. From the point of view of someone on the Moon's surface, a lunation begins with the Sun rising over the portion of the Moon that contains Mare Crisium. Until recently, this was known as the Moon's western limb. It is now known as the eastern limb to avoid any confusion that might arise if people persisted with the original convention that the Sun rises in the east on Earth and in the west on the Moon.

Galileo pointed out that, although all the Earth can see half the Moon, half of the Moon can see all of the Earth. In fact, from the Moon, the Earth maintains an almost fixed position in the sky, and someone on the Moon would always see the Earth in the same position relative to the lunar horizon. There is no earthrise on the Moon. However, since the Earth spins on its axis much more rapidly than the Moon, from the Moon, one can see the entire globe in 24 hours.

The remarkable coincidence between the periods of the Moon's rotation and revolution is due to the tidal effects of the mutual gravitational attraction between the Earth and the Moon. As a result of these tidal forces, which have been acting since the Earth–Moon system was first formed over 4 billion years ago, the Earth's rotation is gradually slowing. Eventually, several billion years from now, the length of the day on Earth will be a synodic month, which itself will be longer than it is now. When this happens, the Moon will be visible from only one side of the Earth. This state of affairs already exists elsewhere in the Solar System: Pluto's rate of rotation about its axis is the same as the orbital period of its satellite, Charon.

9.16 Libration

To a casual observer it appears that we always have the same view of the Moon's surface. Careful observation over a period of days can reveal that this is not quite the case. Using binoculars, and making a note of the features visible at the Moon's edge, it is possible to detect that the Moon appears to wobble very slowly and ever so slightly, allowing us to see a little beyond the edge of the hemisphere that faces Earth. This wobble is known as libration, and it is the result of a number of independent motions. Of these the most pronounced are a libration in longitude and a libration in latitude.

Libration in longitude is a consequence of the Moon's elliptical orbit. The velocity of an object travelling in an elliptical orbit is not constant. It is greatest when closest to the body that it orbits (for an object orbiting the Earth this point is known as perigee), and least when furthest away (known as apogee). So, although the Moon's period of rotation is exactly the same as its orbital period (both are 27.32 days), variations in the Moon's orbital velocity mean that sometimes the Moon is rotating more rapidly than it is orbiting Earth, and sometimes more slowly. At perigee, when its orbital velocity is greater than its rotational velocity, the Moon is not turning fast enough to keep up with the rate at which it is moving around the Earth. This allows us to see slightly beyond the eastern limb for a few days after perigee. Maximum libration in this direction occurs about a week after perigee. At apogee, the rate of rotation exceeds the orbital velocity, and so after the Moon has passed apogee we begin to see beyond its western limb. Maximum western libration occurs about a week after apogee. Libration in longitude amounts to about 6° either way of the average limb of the Moon.

Libration in latitude is due to the inclination of the Moon's axis of rotation to its orbital plane. The angle between them is 83.5°. And, like the Earth's axis, which is itself inclined at 66.5° to its orbital plane, the Moon's axis always points in the same direction in space. The slight difference in

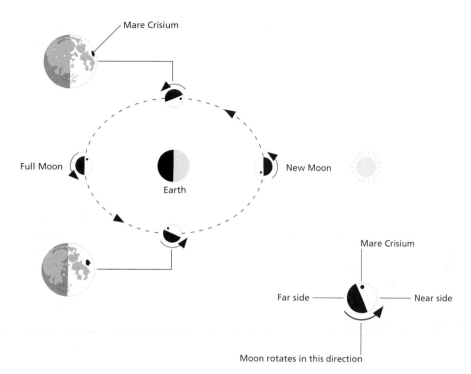

Figure 9.17 Libration. The rate at which the Moon rotates about its axis remains constant, while the speed at which it travels along its orbit varies, being least when it is closest to the Earth. As a result, the Moon doesn't keep exactly the same face pointing at the Earth and so it is possible to see more than half the Moon's surface over one lunation. For example, as the diagram shows, Mare Crisium will seems further from the eastern limb at one point in the Moon's orbit than at another. This phenomenon is known as libration in longitude.

their axial alignments allows an observer on Earth to see alternately beyond the Moon's north and south poles in the course of a lunation. Incidentally, seen from the Sun, the Earth also librates in latitude. As a consequence, the Sun's rays strike each hemisphere more steeply in summer than in winter and polar regions are alternately lit up and plunged into darkness for months at a time.

The net result of these librations is that it is possible to see almost 60% of the Moon's surface from the Earth. However, there are several independent motions that contribute to libration, and it takes many years for all the areas that can be revealed by libration to come into view. The many causes of libration act simultaneously, and as a consequence the point on the Moon's limb showing the greatest amount of libration moves around the edge of the Moon in an anti-clockwise direction during a lunation. Most of the motions that contribute to libration are not easily detected by the casual observer and so they are not discussed here.

To see the effect of the librations in longitude and latitude, it is probably best to begin by finding out which limb of the Moon shows maximum libration on a particular date. The necessary information can be found in the magazine *Sky and Telescope*, or in the handbook of the British Astronomical Association. Mare Crisium, on the Moon's eastern limb, is usually chosen to show libration in longitude. Keeping in mind that this limb points towards the Sun during the first half of the lunation, note that, at extreme western libration, Mare Crisium will be seen as a dark, elongated patch on the edge of the eastern limb of the Moon. A couple of weeks later, at extreme eastern libration, when the Moon has swung in the opposite direction, there is an easily perceptible gap between a more oval Mare Crisium and the eastern limb. Note that the eastern limb of the Moon cannot be seen after full Moon and so, if you use Mare Crisium as an indicator of libration, you must begin to observe it soon after new Moon. If you possess particularly sharp eyes, once you know what to look for, you should be able to detect libration with the naked eye. During the second half of a lunation, the crater Grimaldi, a dark elongated patch near the western limb, can be used as an indicator of libration. Mare Frigoris, near the Moon's north pole, is a suitable feature to use to detect libration in latitude.

9.17 Lunar puzzles

The fact that we observe the Moon from the surface of another world that is itself in motion makes the Moon behave in ways that can seem puzzling. Here are two such puzzles.

The first of these is the way in which the Moon appears to rotate as it moves across the sky. An observer at mid latitudes sees the full Moon rise in the east with its north–south axis inclined to the horizon. As it moves across the sky, the Moon appears to rotate in a clockwise direction so that it crosses the southern sky with this axis approximately perpendicular to the horizon. Finally, when it is on the point of setting on the western horizon, the north–south axis is once again inclined to the horizon. The effect is particularly noticeable at quarter Moon because the asymmetry of its semicircular outline makes it easy to notice the apparent change in orientation. Of course, the Moon cannot actually swing about in this way.

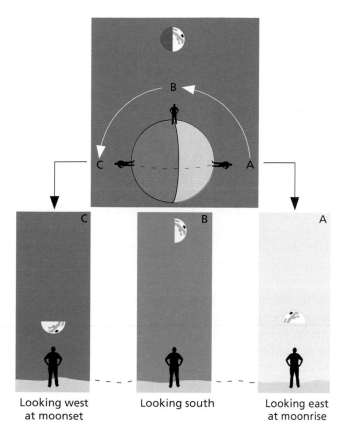

Looking west   Looking south   Looking east
at moonset   at moonrise

Figure 9.18 Moon paradox 1. Because of the Earth's rotation, the Moon appears to rotate as shown in these diagrams. At A an observer will see the Moon rise above the eastern horizon. Its N–S axis is then more or less parallel to the horizon. some six hours later, the observer is at B and the Moon's N–S axis is more or less perpendicular to the horizon. Finally, as the Moon is about to set, its N–S axis is once again parallel to the horizon, but pointing in the opposite direction relative to the observer.

Like the Earth's axis, the Moon's axis always points in the same direction in space.

The illusion is brought about by the gradual change in the orientation of the observer's horizon with respect to the Moon's axis. This change is due to the fact that we are standing on a spinning, spherical Earth. Because of its rotation, we are carried past the Moon in a great arc. The effect this has upon how we see the Moon is shown in figure 9.18. The degree to which the Moon's axis nods back and forth varies with the time of year and with latitude. Someone in the southern hemisphere sees the Moon the other way around, i.e. the Moon's northern pole is closest to the horizon. The effect is further proof that the Earth is a sphere.

A second puzzle concerns the direction from which the Moon appears to be illuminated. This coincides with the Sun only at the beginning or at the end of the lunar cycle, i.e. when the Moon is a slim crescent. Throughout

Figure 9.19 Moon paradox 2. Although the Moon's crescent appears to point away from the Sun, you will realise that this is an illusion by stretching a length of string so that it joins the centre of the Moon to the centre of the Sun. The string will be lined up with the ecliptic.

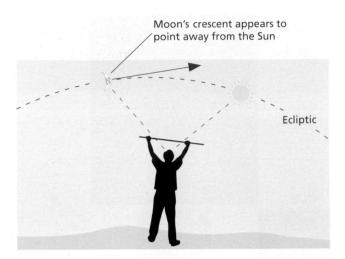

Moon's crescent appears to point away from the Sun

Ecliptic

the rest of the cycle the illuminated portion does not appear to point at the Sun. The effect is very pronounced when one can see either a quarter or gibbous Moon in the sky at the same time as the Sun.

This illusion, for it is an illusion, involves an error of parallax and can be explained as follows. The distance between the Earth and the Sun is approximately 400 times greater than that between the Earth and the Moon. Consequently the direction from which the Moon is illuminated is almost parallel (to within a few minutes of arc) to that from which the Earth is illuminated. This means that the line-of-sight of an Earth-bound observer looking towards the Sun can be taken to be parallel to the direction from which the Moon is illuminated. From the vantage point of someone on the Earth's surface, the two directions do not, indeed cannot, appear to converge and so the illuminated portion of the Moon appears to point away from the Sun. The reason why the effect is less pronounced when the Moon is a crescent is that it is then seen close to the Sun and so the direction from which it is illuminated almost coincides with the Earth observer's line-of-sight to the Sun.

Nevertheless, the fact that the Moon is illuminated by the same body as we are can be confirmed if a long, narrow, straight stick held at arms length is used to join the centre of the visible portion of the Moon to the Sun. The stick will trace out the great circle that joins the two bodies on the celestial sphere. This great circle is the ecliptic. Since each of us is at the centre of the celestial sphere, the projection of any section of a great circle appears as a straight line.

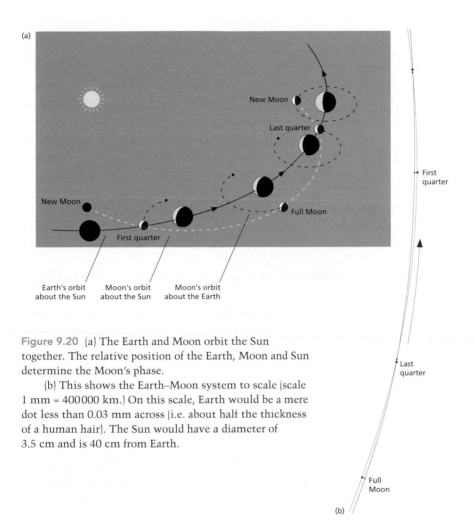

Figure 9.20 (a) The Earth and Moon orbit the Sun together. The relative position of the Earth, Moon and Sun determine the Moon's phase.

(b) This shows the Earth–Moon system to scale (scale 1 mm = 400 000 km.) On this scale, Earth would be a mere dot less than 0.03 mm across (i.e. about half the thickness of a human hair). The Sun would have a diameter of 3.5 cm and is 40 cm from Earth.

9.18 The Moon's phases

At any given moment, half the Moon's surface is illuminated by the Sun. However, the side of the Moon that always faces the Earth, the 'near side', becomes visible to us on Earth in stages from one day to the next because of the Moon's orbital motion about the Earth. These stages are, of course, the phases of the Moon.

These phases are direct proof that the Moon is spherical. No other shape could account for the same sequence, from crescent to full Moon and back again, that we observe in lunation after lunation. The sequence would be quite different if the Moon were, for example, a flat disc, or a cylinder, or even if it were an oval.

The precise moment at which the Moon reaches a particular phase is determined by the relative positions of the Sun, the Earth and the Moon. It doesn't depend on the observer's geographical location. Thus new Moon does not necessarily occur when you see the Sun. The Moon may pass between the Earth and the Sun at the moment when the side of the Earth on which you live is facing away from the Sun. In other words, new Moon may occur at midnight by your local time. Similarly, the moment of the full Moon may coincide with midday, measured by your local time. In this case you would not witness the exact moment of this event, even though you see what appears to be a full Moon rising at the eastern horizon just as the Sun sets in the west.

The point directly beneath the Moon at the moment of full Moon is called the sublunar point. Knowledge of the time at which the Moon is new or full is useful if you want to gauge the likelihood of witnessing a lunar eclipse. The exact time at which the Moon reaches a particular phase can be found in many newspapers, almanacs, the internet or by running a simulation of the sky on a PC.

Visibility of the Moon's phases

	New Moon	Young Moon	First quarter	Full Moon	Last quarter	Old Moon
Position in the sky	Next to the Sun	East of the Sun	90° east of the Sun	Opposite the Sun	90° west of the Sun	West of the Sun
Rises at:	Dawn	Mid-morning	Noon	Sunset	Midnight	Before sunrise
Culminates at:	Midday	Mid-afternoon	Sunset	Midnight	Sunrise	Mid-morning
Sets at:	Sunset	After sunset	Midnight	Dawn	Noon	Mid-afternoon
Times of visibility	Invisible	Evening	Late afternoon and evening	All night	Second half of the night and early morning	Early morning

Since the Sun moves along the ecliptic at a rate of approximately 1° per day, in the course of a lunation it moves approximately 30°. Hence the point on the ecliptic at which the Moon reaches each phase also moves around the ecliptic by approximately 30° from one lunation to the next. One very obvious consequence of this is that the daily path across the sky of each phase is not the same from one lunation to the next.

New Moon

New Moon occurs at the moment when the Moon passes directly between the Sun and the Earth. Such an alignment is known as a syzygy. The Moon is then in conjunction with the Sun and cannot be seen because its illuminated side is pointing directly away from the Earth. The reason why the Sun is not eclipsed at every new Moon is because of the 5° inclination of the Moon's orbit to the plane of the ecliptic. The Moon usually passes slightly above or slightly below the Sun's disc, and so eclipses of the Sun are the exception rather than the rule. See section 12.2 on eclipses of the Sun for more details about the Moon's orbit.

Because of the angle between the Moon's orbital plane and the ecliptic the angular distance between Moon and Sun at new Moon can be as much as 5° and so you might expect that there would be times when the new Moon is visible as a slender crescent. The reason it never is, is explained below.

Young Moon

It is only by a combination of experience and luck that you will see the first sliver of a crescent Moon less than 30 hours after the new Moon. The record for such an observation appears to stand at 14.5 hours, though this was achieved with a telescope. Most people first catch sight of a young Moon, which is often incorrectly called a new Moon, above the western horizon on the evening of the second or third day after new Moon. It will then be between 24° and 36° from the Sun. Knowledge of this angular separation can help you find a young Moon. 24° is a little bit more than the angle subtended by a full handspan held at arms length. Binoculars can be a great help because the narrow sliver of a young Moon is difficult to see, partly because it is intrinsically faint and seen against the twilight sky, and partly because it is so narrow. If you are in the northern hemisphere, your best chance of an early sighting of a young Moon will always occur at or around the spring equinox (see section 8.4 for the reasons).

The circumstances that determine the earliest sighting of a young Moon were investigated in detail by André Danjon, a French astronomer. He concluded that it is not possible to see a crescent when the Moon is less than 7° from the Sun. There are several reasons for this. The cycle of lunar phases begins with sunrise over the eastern horizon of the near side of the Moon, which means that the segment of the Moon's surface that is directly visible from Earth is illuminated by a low Sun. If the Moon were perfectly smooth, this segment would be visible shortly after new Moon as a narrow, fully semicircular crescent, though it would not be very bright because the lunar surface has such a low albedo. But, as we first mentioned in section 9.7, the Moon's surface is uneven, and the illuminated slopes of hills within the segment of the crescent mostly face away from the Earth. Hence they are not seen. The reverse slopes of these hills, which do point towards Earth, lie within the hills' own shadow. Shadows on the Moon are, of course, pitch dark because there is no atmosphere to scatter light into them. Self-shadowing thus renders these reverse slopes completely invisible to an observer on Earth. At the same time, when the Sun is very low, shadows extend far beyond the foot of the reverse slopes of hills, hiding even more of the lunar surface from the Earth-bound observer. The net effect of all this is to reduce the apparent width and brightness of the crescent below that necessary for it to be visible from Earth until the Moon reaches an elongation that is at least 7°. Self-shadowing is the reason why the new Moon is never visible even when it is at its maximum angular distance of 5° from the Sun above or below the ecliptic.

In the course of his observations, Danjon also noted that the tips of the horns of a crescent Moon are not visible when the Moon is less than 40° from the Sun. 40° is approximately the elongation of the Moon three days after new Moon. Hence the arc of the crescent does not extend to a full semicircle until the Moon is some three to four days old. The invisibility of the horns is also due to self-shadowing, and this adds to the difficulty of seeing the young Moon. Danjon once observed the waning Moon with an elongation of only 8°, i.e. about 16 hours before new Moon, but he needed binoculars to see it. On that occasion, the crescent was only 4 arc seconds wide and extended over only 60°. We are so used to drawings incorrectly showing a crescent extending to a full semicircle, that we don't notice that it doesn't do so until the third or fourth day of a lunation. It isn't difficult to notice when the crescent is less than a semicircle: hold up a straight edge, such as a pencil, so that it joins the tips of the crescent. You can tell right away whether or not the tips lie on opposite ends of a semicircle.

First quarter

It takes approximately 7.5 days for the Moon to reach first quarter. The Moon is now 90° east of the Sun and rises approximately 6 hours after sunrise. This phase takes its name from the fact that it marks the point at which the Moon has travelled one quarter of its total orbit. It is a good time to look at the Moon through binoculars because you can see much of the heavily cratered lower southern hemisphere in strong relief.

Gibbous Moon

The original meaning of gibbous is 'hump backed'. The term is applied to the Moon when it is either between first quarter and full Moon (when it is getting larger, or waxing) or between full Moon and last quarter (when it is getting smaller, or waning).

Full Moon

Full Moon occurs exactly half way through the synodic month, 14.75 days after the new Moon. The Moon is now 180° to the east of the Sun and so moonrise occurs 12 hours after sunrise. This is why the Moon and the Sun usually can't be seen in the sky together. An exception to this rule occurs occasionally when the exact moment of full Moon coincides with the moonrise for a particular location. In such circumstances, atmospheric refraction may be great enough to lift the images of either the Sun or the Moon above the horizon so that both can seen simultaneously for a few minutes.

To the naked eye, the disc of a full Moon is indistinguishable from a gibbous Moon that is almost full because the rate at which the size of the lunar disc changes from day to day is least around full Moon. Nevertheless the difference in brightness between them is noticeable to the naked eye (see section 9.7 for more on the Moon's brightness.)

Lunar eclipses can occur only at full Moon, through not at every full Moon. See section 10.7 for information on lunar eclipses.

Last quarter

Last quarter occurs on the 22nd day of the synodic month. On this day the Moon rises 18 hours after the Sun, i.e. around midnight. At sunrise, the Moon is half way across the sky and remains visible in the morning sky until midday, by which time it has reached the western horizon, where it sets.

The last quarter is darker than the first quarter because it is largely covered by Oceanus Procellarum and Mare Imbrium.

You can make use of the last quarter to work out the direction in which the Earth is moving through space. If you look at figure 9.20, which shows the Moon's motion around the Earth, you can see that the Earth is itself moving towards the point occupied by the last quarter Moon. It takes approximately 3.5 hours for the Earth to reach this point. Of course, by the time it is reached, the Moon is no longer there because it too is orbiting the Sun along with the Earth.

Another interesting fact that can be deduced from figure 9.20 is that the Moon moves a greater distance in its orbit about the Sun between first quarter and last quarter than it does between last quarter and first quarter. This is because of its motion about the Earth: between first and last quarter it is moving in roughly the same direction as the Earth, whereas between last and first quarter it is moving in the opposite direction. Hence its relative velocity around the Sun is greatest at full Moon and least at new Moon.

Waning crescent

The crescent of the old Moon rises at the eastern horizon just ahead of the Sun. In appearance it is a mirror image of the waxing crescent of the young Moon and its unilluminated portion may be lit up by earthlight. This phase is most prominent in the northern hemisphere at the autumn equinox when it is high above the horizon before sunrise.

Chapter 10 | ECLIPSES

> Now eclipses are elusive and provoking things . . . visiting the
> same locality only once in centuries. Consequently, it will not
> do to sit down quietly at home and wait for one to come, but a
> person must be up and doing and on the chase.
>
> Rebecca R. Joslin, *Chasing Eclipses: The Total Eclipses of 1905,*
> *1912, 1925*, Walton Advertising and Printing, 1929

10.1 Chasing eclipses

The proverb that lightning never strikes twice in the same place is, in fact,
false. The proverb is truer of an eclipse. It has been calculated that, on
average, a total eclipse of the Sun can be seen from the same spot only once
every 410 years. So, if you really want to see one, you must take Rebecca
Joslin's advice and go '. . . on the chase'. If you set your heart on seeing one,
sooner or later, you will see an eclipse; and you may find, as many have
before you, that no sooner have you seen it, than you start planning to see
the next one. Part of the attraction is that solar eclipses don't occur very
often. On average there is one every 18 months. But, even if they were
more common, people would still go out of their way to see them: they are
spectacular events, arguably the most breathtaking that the sky has to
offer.

An eclipse is also the most tangible of all astronomical events. What
makes it so is that, during an eclipse, one celestial body casts its shadow on
another. During a lunar eclipse, as the Moon becomes enveloped in the
Earth's shadow, it seems as if the vast space between them has been bridged.
There are times during a lunar eclipse that I have felt I was as close to touching the Moon as I shall ever get without actually walking on it. A total solar
eclipse is even more palpable since it affects the very place from which it is
seen. Unlike the Earth's shadow, which becomes visible only when it falls
on the Moon, the Moon's shadow can sometimes be seen in the sky just
before and after totality.

Eclipses occur because of a remarkable coincidence. Although the
diameter of the Moon is some 400 times less than that of the Sun, the
Moon is 400 times closer to us, and so both the Sun and Moon have a
similar apparent size seen from the Earth. This allows the Moon to completely cover the Sun when the Earth, Moon and Sun are aligned. Such an

alignment is known as a syzygy. Although syzygies happen twice in every lunation, at new Moon and full Moon, or at least 24 times a year, eclipses are rare because the Moon's orbit is slightly inclined to that of the Earth. Seen from the Earth, in most syzygies the Moon passes either above or below the Sun.

There are two types of eclipse: a solar eclipse, which occurs when the Moon passes directly between the Earth and the Sun at new Moon, and a lunar eclipse, which occurs when the Earth is between the Moon and the Sun at full Moon. Strictly speaking in both cases it is the Sun that is eclipsed. If you were on the Moon during a lunar eclipse, you would see the Sun disappear behind the Earth. But, since we are on Earth, it is the Moon that is seen to disappear, or at least grow dim, and the event is rather misleadingly known as an eclipse of the Moon.

10.2 Solar eclipses

Much is made of the coincidence between the apparent size of the Sun and Moon. On closer examination this turns out not to be quite true. The Earth's orbit around the Sun is almost circular, which means that the Sun's apparent diameter of half a degree remains more-or-less the same throughout the year. On the other hand, the Moon's orbit around the Earth is markedly elliptical, and so its apparent diameter changes significantly compared with that of the Sun. In fact, the difference in their angular diameters, small though it is, makes possible two types of eclipse: a total eclipse, in which the Sun's disc is completely covered by the Moon's disc for a short while, and an annular eclipse in which the edge of the Sun remains visible throughout the eclipse. Annular eclipses are more frequent than total eclipses because the average distance between the Earth and the Moon is such that the average angular diameter of the Moon is very slightly less than that of the Sun. Hence, in most eclipses, the Moon's disc is not quite large enough to cover the Sun's disc completely.

The relative sizes of the solar and lunar disc also affect the duration of an eclipse, which in turn affects many of the phenomena that occur during an eclipse. When the Moon's apparent size is as large as it can get relative to that of the Sun, totality can last up to seven and a half minutes, and when it is as small as it can get, an annular eclipse can last just over 12 minutes. However, most eclipses last for less than this because they seldom happen when the Moon is either extremely close to or far from Earth.

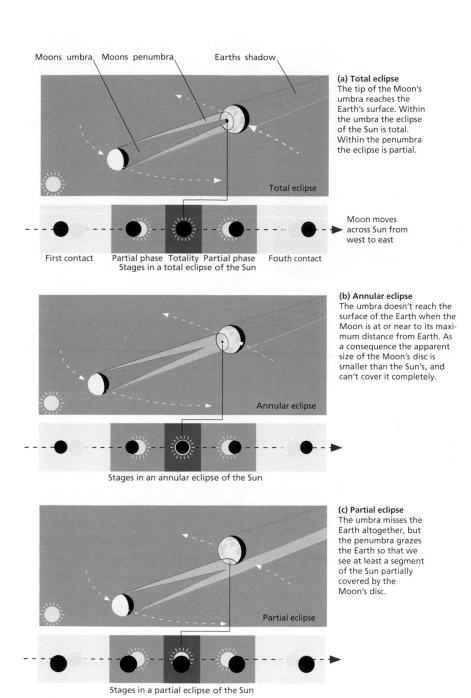

Moons umbra   Moons penumbra   Earths shadow

(a) Total eclipse
The tip of the Moon's umbra reaches the Earth's surface. Within the umbra the eclipse of the Sun is total. Within the penumbra the eclipse is partial.

Total eclipse

Moon moves across Sun from west to east

First contact   Partial phase   Totality   Partial phase   Fouth contact
Stages in a total eclipse of the Sun

(b) Annular eclipse
The umbra doesn't reach the surface of the Earth when the Moon is at or near to its maximum distance from Earth. As a consequence the apparent size of the Moon's disc is smaller than the Sun's, and can't cover it completely.

Annular eclipse

Stages in an annular eclipse of the Sun

(c) Partial eclipse
The umbra misses the Earth altogether, but the penumbra grazes the Earth so that we see at least a segment of the Sun partially covered by the Moon's disc.

Partial eclipse

Stages in a partial eclipse of the Sun

Figure 10.1 Solar eclipse. There are three types of solar eclipse: (a) total, (b) annular, and (c) partial. In each case, the Moon passes between the Sun and the Earth.

Figure 10.2 During totality the Sun's faint atmosphere, or corona, becomes visible. The first photograph shows the solar corona seen during the eclipse of 1991 from La Paz, Mexico. To reveal the detail of the complicated filament structure of the corona the photograph was taken using a special telescope which compensates for the wide range of brightness of the corona. A few pink prominences are seen near the limb. These belong to a lower layer in the solar atmosphere called the chromosphere. (*Photo* Dr Francisco Diego)

The second photo shows the solar corona seen during the eclipse of 1980 from the Tsavo West National Park in Kenya. The corona appears quite uniform around the Sun as this eclipse occurred when solar activity was at a maximum in its 11-year-long cycle. (*Photo* Dr Francisco Diego)

A third type of eclipse occurs when the Moon's umbral shadow just misses the Earth. This is the most common type of eclipse. The most that you will see in such circumstances is a partial eclipse of the Sun. Figure 10.1 shows the relative positions of Earth and Moon that give rise to each of the three types of eclipse.

10.3 Preparing to see an eclipse

There is much more to a solar eclipse than the Sun being briefly covered by the Moon. A total eclipse of the Sun is a light show on a grand scale, and it's worth going out of your way to see one. But to fully enjoy the spectacle you should do some homework beforehand. An eclipse is accompanied by many subtle effects, and these are easily missed unless you know what to look for.

Information about future eclipses, weather prospects, duration of totality, and many other things is always available well in advance of any eclipse on several web sites (see Further reading).

It's not just a visual spectacle that awaits you. Many people claim to be elated by an eclipse, though I personally find eclipses are ominous rather than exhilarating. I'm not alone in this: a common response to eclipses among primitive folk was fear, and some would even hide away during totality. Once you've seen a total eclipse it isn't difficult to imagine its effect on superstitious people who knew nothing of it: the Sun vanishes for several minutes without much warning, spreading fear and confusion. Nevertheless, don't imagine that a total eclipse is a show-stopper: I've seen drivers switch on their headlights, and continue driving during totality. And there are plenty of people who, having seen an eclipse, wonder what all the fuss is about.

Perhaps the most important thing is to decide where you will get the best view of the eclipse. Begin by finding out the path of totality. This is the track of the Moon's umbra across the Earth's surface. You will be limited by the fact that 70% of the Earth's surface is covered by water, so only a fraction of any given eclipse is usually visible from dry land. Next check the weather prospects at different places. It is somewhat ironic that, although an eclipse can be predicted to the second, hundreds of years in advance, we can seldom be certain of the weather from one day to the next. Avoid anywhere that cloudy conditions are usual. Clouds can appear at the last moment and prevent you seeing the eclipse, though the sky will still grow dark during totality.

Another thing to consider is the duration of totality, or how long the Sun is going to be covered by the Moon. This varies along the path of totality. Usually it is least at the beginning and end of the path of totality, and greatest at the midpoint. Even if you don't want to head for the point where totality lasts longest, it's worth selecting somewhere that is as close as possible to the centre of the Moon's umbra. Totality will always last longer there than somewhere near the edge of the umbra.

No single eclipse will give you the best view of all the phenomena that occur during an eclipse. The darkness of the sky during totality depends on how much larger the Moon's disc is than the Sun's, and how close you are to the centre of the moon's umbra. The larger the Moon, and the closer you are to the centre of its shadow, the darker the sky will be. A dark sky means that you will see many more stars and planets, and that you will get a better view of the fainter parts of the corona. However, Baily's beads, the diamond ring and prominences (see section 10.5), are usually more obvious when the Moon's disc is only just able to cover the Sun.

Safe viewing is another thing that you must get sorted in good time. Don't wait till the last minute and improvise. You probably already know that you should always take the greatest care not to look directly at the Sun. Normally, of course, we avoid doing this. But during the partial phase of an eclipse, you will want to look at it to see what's going on. This is why so many people injure their eyes, sometimes permanently, during an eclipse. So it's really vital that you decide well in advance how you are going to look at it safely. There are several ways in which you can do this. You can use pinhole projection, telescopic projection, or wear suitable filters that reduce the intensity of the Sun's light while absorbing its infrared and ultraviolet rays. Try them all out to see which you prefer. You may find that you want to use all of them on the day, if only because it helps to while away the time during the partial phase. Don't look at reflections in glass or water, and don't look through smoked glass, or sun glasses, or use any of the other methods prescribed by folklore. You can't be sure that they will reduce all the Sun's radiations to safe levels. Always remember: don't look at the Sun directly until totality. However, it is perfectly safe to look at it directly while it is covered by the Moon.

An eclipse is probably the only time when you look at the sun directly, albeit with eye protection. You will probably be struck by how small it is. Projecting its image with a pinhole won't allow you to notice this. Even if you know that the Sun's disc is the same size as that of the Moon, without eye protection, it's so bright that its glare makes it seem much larger than it really is.

A final point: is it worth photographing or videoing an eclipse? I think that you shouldn't try to preserve your first eclipse for posterity. At least, avoid taking photos or using your video camera during totality. Totality doesn't last long. It's not easy to take good photos of the corona, and time spent adjusting your camera is time you should spend enjoying the spectacle. If you want photos, you can always buy them from professionals later, or take them yourself during your next eclipse.

10.4 Watching a solar eclipse

From an astronomical point of view, an eclipse is the result of the relative positions and relative sizes of the Earth, Moon and Sun. But, for most of us, the attraction is not the astronomical event, the syzygy, it's the spectacle in the sky.

In an eclipse, the Moon's shadow overtakes the Earth and races across the Earth's surface. Since the Earth spins in the same direction as that in which the Moon orbits it, the ground speed of the shadow is less than it would be if the Earth were stationary, and it depends on latitude, being least at the equator. To give you a rough idea, the ground speed of the shadow is approximately 500 m/s within the tropics, rising to about 1000 m/s at the poles. Consequently, the longest lasting eclipses are seen from the tropics.

The average width of the whole shadow, from one edge of the penumbra to the other, is about 7000 kilometres. But to see a total eclipse you need to be within the central portion of the eclipse shadow, the umbra, which is approximately 200 km wide. Even with a ground speed of 1000 m/s, it usually takes several hours for the shadow to move across the Earth's surface. Anyone within the path of the umbral shadow, known as the path of totality, will see the Sun completely eclipsed for however long it takes the umbra to overtake them.

Seen from anywhere along the path of totality, the Moon takes between two and three hours to go right across the face of the Sun. But you won't notice much change in your surroundings, or in the sky, until about a quarter of an hour before totality. After that, things change quickly.

Although there isn't much to see until those last few minutes, it's worth arriving at the spot from which you want to see the eclipse an hour or two before totality. This will give you time to get your bearings, sort out your equipment, and so on. A few hours earlier, far to the west, the Moon's shadow caught up with the Earth, and is now rushing towards you across land and sea. At the same time the Earth's spin carries you eastwards away from the Moon's shadow. But the Moon's shadow moves more swiftly, and soon catches up with you. As it overtakes you, you see the Sun gradually disappear behind the Moon. The Sun vanishes for the few minutes it takes the umbra to pass over you.

Unless you are at the western extremity of the path of totality, the point at which the Moon's umbra first falls upon the Earth's surface, you won't be the first to see the eclipse. If you are several thousand kilometres to the east of that point, people to the west of you will have already seen totality when you first notice that the Moon's disc has begun to encroach on the Sun, a moment known as first contact. You are now on the eastern edge of the penumbral shadow. The centre of the shadow, the umbra, is some 3500 kilometres to the west, and, weather permitting, people there are already seeing the total eclipse.

For about half an hour after first contact there is little obvious sign that the Moon now overlaps the Sun. Unless you look at the Sun, using a safe viewing method, you can't see that the Moon is taking an ever-increasing

Figure 10.3 The sky during totality seen from the Bolivian altiplano during the total phase of the solar eclipse of 1994. Near the eclipsed Sun are the bright planets Venus (top right) and Jupiter (bottom). The horizon during totality is reddened. (*Photo* Ros Brown)

bite out of the Sun. It's only when the Sun is more than half covered that you begin to be aware that it's noticeably less warm, and the sky and your surroundings have darkened slightly.

If you have a clear view to the horizon, in the last moments before the Sun disappears behind the Moon you may see the Moon's shadow racing towards you looking like a vast, silent thunderstorm.

The transition to totality takes place so quickly and is so abrupt that it may take you by surprise. As the Sun vanishes, its narrow crescent breaks up into several bright points of light. These are known as Baily's beads, and are so brief that if you are not prepared for them you can easily miss them. The whole sky suddenly darkens to a deep twilight, and the horizon turns orange. It's as if someone had used a dimmer switch to turn the Sun off: there is the same abrupt change from dim illumination to darkness.

The second contact is totality. The Sun has been replaced by a black disc, which is the unilluminated Moon, now the darkest thing in the sky.

Surrounding the dark disc is the irregular pearly glow of the Sun's corona, about as bright as a full Moon.

All too soon you become vaguely aware that the sky is growing brighter in the west. The third contact is when the Sun suddenly reappears on the western side of the Moon's disc, often flaring out as a single flash of light that has come to be called the 'diamond ring' effect, immediately followed by Baily's beads. It is time to replace your eye protection. You are now at the western edge of the Moon's umbra. Totality is over for you. Your watch tells you it lasted several minutes, but it seems like only a few seconds.

The return to a partial eclipse is an anticlimax. The sky seems rapidly to return to normal, like a movie being run backwards; but the high point is over. It seems pointless to continue looking at the sky, or the ground for a last look at the things that were so fascinating merely a few minutes earlier. It's time to compare notes with others, and talk of the next eclipse. You may still be talking at fourth contact, an hour or more later, which is when the Moon's disc just ceases to cover any part of the Sun.

And what if you are not on the path of totality? The width of the Moon's partial shadow, or penumbra, is about 3500 km either side of the region of totality. Anyone within the penumbra will see a partial eclipse. The fraction of the Sun that is covered by the Moon decreases the further away you are from the central region of totality. If you are within a couple of hundred kilometres of the path of totality, you will notice the strange metallic light, the Sun will feel much less warm, shadows will grow sharp, and you will see crescents under trees.

10.5 Eclipse checklist

Here are some things to look out for during a total eclipse.

First contact

This is the moment at which the edge of Moon's disc first appears to make contact with the edge of the Sun's disc, so that they seem to be just touching. It takes place about one hour before totality, and marks the start of the eclipse as seen from your position on the Earth's surface. You are now on the eastern edge of the Moon's shadow. However, people to the west of you are already within the shadow, and first contact for you occurs at the same time as totality for people some 3600 km to the west of where you are. You won't notice any change in sky, or in your surroundings, for at least half an hour after first contact, though you will be able to follow the Moon's progress across the Sun's disc moment by moment if you view them through a filter.

Shadows

During the partial phase, as the Moon's disc gradually creeps across the Sun, eventually reducing it to a thin crescent, shadows lose their penumbra and become sharper. Just before totality, the shadows of even the smallest things become visible: you can see individual hairs on the shadow of your head. The patches of light in the shadows cast by trees and bushes in leaf change from their characteristic oval shape to that of a crescent. During an annular eclipse, these patches are ring-shaped. If there are no trees around, spread all your fingers apart, place one hand diagonally across the other, and use the resulting mesh of fingers to cast shadows on the ground. It works better than leaves.

Airlight

About a quarter of an hour before totality, the sky has darkened noticeably and the Sun feels much cooler. Everything around you is bathed in a wintery, metallic blue light. This is airlight, which, of course, is distinctly blue.

Shadow bands

Just before, and just after totality, when the Sun's disc is no more than a thin sliver, faint parallel bands of light and dark may be seen moving over the ground. These are known as shadow bands, and are due to the scintillation of the Sun's narrow crescent brought about by atmospheric turbulence. Turbulence causes the sunlight to be concentrated in bands that are parallel to the Sun's crescent. The spaces between the bands, being less well illuminated, look like long shadows. Shadow bands stand out particularly well when they are formed on snow.

Baily's beads

Just before the narrow crescent completely disappears behind the Moon, it breaks up into several bright segments. These are known as Baily's beads, and are due to the sunlight streaming through valleys and gaps between the mountains on the rim of the Moon's disc. If you are on the edge of the path of totality, you may see Baily's beads work along the edge of the Moon during totality.

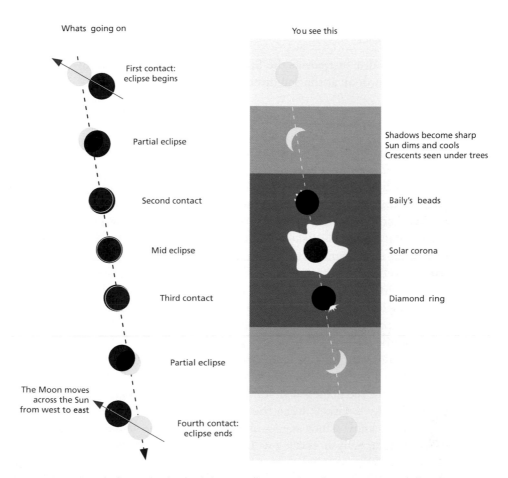

Figure 10.4 Stages of a total eclipse of the Sun, showing the relative positions of the Sun and Moon and what you can expect to see from the ground. In this diagram, the Sun (the yellow disc) is made slightly smaller than the Moon (the black disc) so that you can clearly see the relative position of Sun and Moon from first contact through to last contact.

The likelihood of doing so is greatest if you are on the southern edge of the Moon's umbra because the Moon's surface is rougher along its southern limb.

A phenomenon associated with Baily's beads is the so-called 'diamond ring' effect. This is the name give to the last flash of light that precedes totality, or the first one at the end of totality.

Both Baily's beads and the diamond ring effect are never the same from one eclipse to another because the profile of the edge of the Moon through which the last rays of the Sun are seen changes from one eclipse to the next.

Second contact

The Moon's disc just covers the Sun. This marks the beginning of totality. You may now look at the Sun directly without protection.

The sky during totality

With the onset of totality, the sky darkens to twilight. It can never become as dark as it is at night because the Moon's shadow only covers a small area of the Earth. Airlight from beyond the Moon's umbra illuminates the atmosphere within it, turning most of the sky a deep twilight blue. The horizon reddens because of selective scattering of airlight from the horizon beyond the umbra. Horizon airlight is usually white, but when you look at it from within the umbra you see the larger, unscattered wavelengths. These appear reddish to the eye. The absence of direct sunlight within the umbra means that the blue light that is scattered out of the horizon light by the air is not made up within the atmosphere directly around you since there is no further light source. A scaled-down version is noticeable in distant clouds during daylight: they can also look faintly pink.

Depending on how dark the sky becomes, and it varies in brightness from one eclipse to another because of the relative sizes of the Moon and Sun, you will see planets and stars, as many as you might see during twilight. The sky is darkest when the Moon's apparent size is at its greatest. If you want to know which stars and planets should be visible during totality in a particular eclipse, check a web site to find out.

The solar corona

During totality, the pearly aureole of the Sun's corona can be seen around the eclipsed Sun. This is the atmosphere of extremely hot gas that surrounds the Sun and which is usually invisible because of the brightness of the sky around the Sun. It is about half as bright as the full Moon and so does not appreciably illuminate our surroundings during totality. Its shape varies from one eclipse to another, and is determined by the Sun's activity, which reaches a maximum every 11 years. When the Sun's activity is at its greatest, the corona spreads equally in all directions; when the Sun is less active, the shape of the corona is determined by the Sun's magnetic field. It does not spread out equally but is confined to particular directions. The solar corona seen during an eclipse should not be confused with the corona that is seen when the Sun is partially obscured by thin clouds. You may, of course, see such coronas during a partial eclipse, particularly because the sky immediately around the Sun is not bright enough to dazzle you.

Effect on animals

Some animals (though not all) react to totality as if it were the onset of night. Totality doesn't last long enough for the darkness it brings to trigger unusual behaviour. Birds can become disorientated, and insects act as if night is falling. But dogs and other mammals appear not to notice that anything has changed.

Third contact

The Sun just emerges from behind the Moon. Totality is over. The moment is frequently announced by the diamond ring effect, and a succession of Baily's beads. Be ready to replace your eye protection the instant you notice the western edge of the Moon brightening.

Fourth contact

The end of the eclipse about and hour after totality: the Moon's disc no longer covers any part of the Sun. The whole eclipse, from first contact to fourth contact takes about two hours.

10.6 Explaining a solar eclipse

Although an eclipse of the Sun can occur only at new Moon, solar eclipses do not occur every time the Earth and Moon are in syzygy because, as we have seen, the Moon's orbital plane is inclined at an angle of 5° to the plane of the ecliptic. The two points at which its orbital plane crosses the ecliptic are known as nodes. The Sun can be eclipsed by the Moon only if the new Moon occurs when the Moon itself is at, or quite close to, one or other of these nodes. A further complication is that the nodes are not stationary: they too orbit the Earth, or precess, taking approximately 18.6 years to go right around the ecliptic. During this period there are just over 220 lunations. Hence from one lunation to the next, the line that joins the nodes – known as the line of the nodes – does not alter its orientation by very much, and points more or less in the same direction in space. This means that one or other of the nodes lies directly between the Earth and Sun only twice a year.

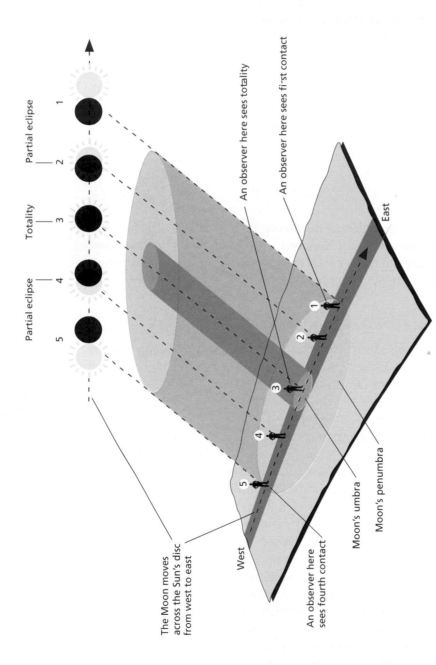

Figure 10.5 This diagram is a snapshot of a moment in a total eclipse of the Sun. It shows what is seen by five observers at different positions along the path of the Moon's shadows across the surface of the Earth.

The eclipse begins when the edge of the penumbra reaches you at the moment of first contact. It takes about another hour for the umbra to reach you. You see totality as the umbra passes over you. The eclipse ends at fourth contact, an hour later at the western edge of the penumbra.

The reason that the Moon does not have to be exactly at a node to eclipse the Sun is that neither the Sun nor the Moon is a point. When the Moon's angular diameter is taken into account, it can be shown that the greatest distance from one of the nodes at which the Moon can just touch the Sun's upper or lower limb is approximately 15°. This distance is known as the eclipse limit. If a new Moon occurs just inside the eclipse limit, it will be seen from Earth to cover a tiny segment of the Sun, and a partial eclipse will result. The closer to the node that a new Moon occurs, the more of the Sun's disc is covered. Total eclipses – when the Moon passes directly in front of the Sun – become possible when a node is within 10° of the Sun.

Eclipse limits can be converted into periods of time because the Sun travels around the celestial sphere at approximately 1° per day. Hence it takes about 30 days to travel from one eclipse limit to the other. This means that there are approximately 15 days either side of an alignment between a node, the Sun and the Earth, during which an eclipse can occur. This period is known as an eclipse season. There are usually two eclipse seasons every year and they occur approximately six months apart. The Moon, of course, takes slightly less than 30 days to orbit the Earth synodically, i.e. from one new Moon to the next, and so two eclipses can occur within the same eclipse season. This, however, can happen only if the first eclipse occurs close to one of the eclipse limits, giving the Moon time to orbit the Earth and catch up with the Sun before the Sun has reached the next eclipse limit. Both these eclipses will be partial.

Given that there are usually two eclipse seasons every year, the maximum number of solar eclipses that we can normally expect in a year is four. These will all be partial. On the other hand, since at least one eclipse must occur during each eclipse season, there are always at least two eclipses per year. Both of these may be total, though they need not be. In some two-eclipse years, both are partial. Exceptions to these rules are due to the precession of the nodes.

As we have seen, the line of the nodes precesses gradually, taking approximately 18.6 years to return to the same orientation in space. This means that it precesses approximately 19° per year. The effect of this is that node alignment takes place every 173 days rather than every 183 days (or six months), which would be the case without precession. The eclipse year, during which there are two node alignments, is thus 346 days long – 19 days shorter than the calendar year. Hence it is sometimes possible for three eclipse seasons to occur within the space of a calendar year if the last eclipse season of the previous year occurs during the last few days of that year. This allows an eclipse to occur before the middle of January of the following year,

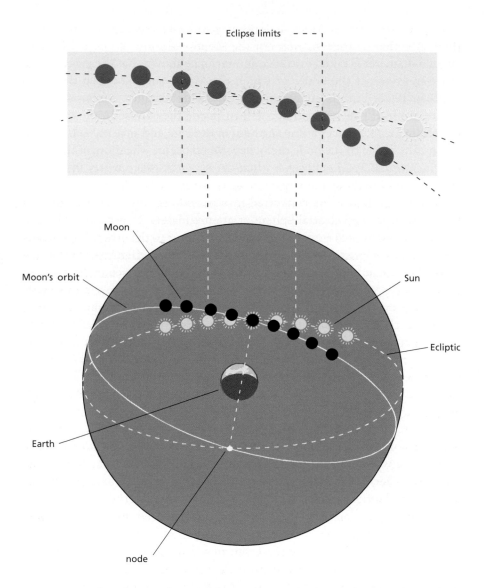

Figure 10.6 Eclipse limits. The Moon does not have to be directly between the Sun and the Earth for it to hide at least some of the Sun.

and leaves time for two further eclipse seasons before the year is out. In these circumstances, five solar eclipses are possible in a year.

Another consequence of the difference between the calendar year and the eclipse year is that the calendar date on which each new eclipse season starts is about 10 days earlier than the one preceding it.

The Moon's orbital motion is fearfully complex, and as a consequence no two successive solar eclipses can be seen from the same location. Indeed it might be several lifetimes before a full total eclipse occurs twice at the same place. On the other hand, since the Sun is eclipsed at least once or twice a year, then if you are really determined to see one, all you have to do is travel to a suitable location. Information on future eclipses is available from several web sites.

Although eclipses have been happening since the Earth and Moon were formed, people have been recording them for, at most, the last 3000 years. Before records were kept, an eclipse of the Sun would always have been viewed as a one-off event, and a hugely unsettling one at that. Since total eclipses of the Sun are seldom visible from the same place more than once in several hundred years, it seems unlikely that people in ancient times could have learned about them from folk histories.

It is often claimed that the Babylonians were the first people to discover how to predict eclipses. This claim is based on a series of misunderstandings that have crept into the history of astronomy, and which it has proved difficult to expunge. The truth is that the Babylonians, despite their deserved reputation for observation and record keeping, were not particularly able astronomers. They had no grand scheme of the heavens which would have enabled them to understand the mechanisms that give rise to eclipses. The honour of discovering how to predict eclipses accurately goes to Hipparchus who lived on the island of Rhodes during the second century B.C. Nevertheless, crude predictions of eclipses, particularly lunar eclipses, were possible before Hipparchus because both Babylonian and Greek astronomers had access to records of astronomical events that covered several centuries, and so were able to spot patterns in the occurrence of both lunar and solar eclipses. The fact that there is a pattern is not in itself particularly remarkable because two or more cyclical events will always come into phase with one another given a sufficient number of cycles. Such coincidences enabled ancient astronomers to lay the foundations of modern astronomy because they suggested that the movements of the heavens were orderly, and so open to rational explanation.

Babylonians had noticed that similar eclipses tend to occur approximately 18 years apart. Eclipses are considered to be similar if both are either total, annular or partial. The reason is that 223 synodic months, the duration of a lunation, are almost the same as 19 eclipse years: 223 synodic months equal 6585.32 days, and 19 eclipse years equal 6585.78 days, a difference of some 11 hours. 6585 days is approximately equal to 18 years and 11 days, a period known as a Saros cycle. The term 'Saros' is a corruption of a Babylonian word for a period of 3600 years, and was unwittingly foisted on modern astronomy by Edmund Halley. The Babylonians themselves were probably unaware of the Saros cycle in the form in which it is known today.

The reason why the 18-year Saros cycle can be used to predict eclipses is that the Moon is in almost exactly the same place in the sky at the end of this period as it was at the beginning. And, if the cycle begins with an eclipse, it is likely to end with one. However, 19 eclipse years are about 11 hours longer than 223 lunations, and so each new Saros begins with the Moon fractionally further west of its position with respect to the Sun than it was at the start of the cycle. This causes the eclipse shadow to fall at a lower latitude than before. The tiny westward shift of the Moon at the end of each 18-year period means that each Saros is itself part of a longer cycle that takes about 1300 years to complete. Each of these longer cycles begins with a partial eclipse visible only from the North Pole. With each succeeding Saros, the eclipse that starts it off crosses the Earth's surface at a slightly lower latitude until, at the end of several hundreds of years there is a final partial eclipse visible only from the South Pole. This marks the end of this particular sequence of Saros cycles. But others are in progress. Indeed, there are some 80 individual Saros cycles at any one time.

10.7 Eclipses of the Moon

A lunar eclipse occurs when the Moon passes through the Earth's shadow. This can only happen at full Moon when the Earth lies between the Moon and the Sun. However, the Earth's shadow is so much wider than the Moon's shadow that anyone who is on the night side of the Earth during the eclipse will be able to see at least some phase of the event. This means that many more people have seen a lunar eclipse than a solar one. Despite this, lunar eclipses are not quite as frequent as solar ones because the width of the Earth's shadow at the point where it crosses the Moon's orbit is less than the width of the eclipse limits within which solar eclipses can occur. In other words, the Moon spends more time in the region where it can eclipse the Sun than it does where it can itself be eclipsed by the Earth.

The width of the Earth's umbra through which the Moon passes is some

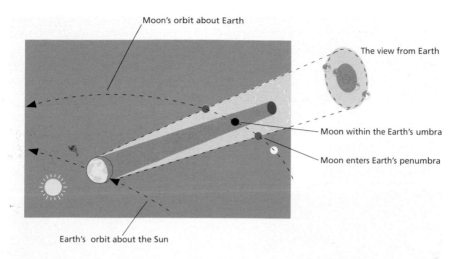

Moon's orbit about Earth

The view from Earth

Moon within the Earth's umbra

Moon enters Earth's penumbra

Earth's orbit about the Sun

Figure 10.7 A lunar eclipse occurs when the Moon passes through the Earth's shadow and is no longer directly illuminated by the Sun.

9200 km. The corresponding penumbra is 16000 km wide. A total eclipse of the Moon can last for up to 1 hour 47 minutes because the umbra is so broad. However, because of the inclination of the Moon's orbit to the plane of the ecliptic, the Moon does not always pass through the centre of the umbra, and therefore three distinct types of lunar eclipse are possible: total, partial and penumbral.

A total lunar eclipse will usually occur during the same lunation as a solar eclipse. The reason for this is that the Moon must be close to one of its orbital nodes for it to be eclipsed, and if one of these nodes lies between the Sun and the Earth, then the other one must lie on the other side of the Earth, i.e. at the point at which full Moon will occur for that lunation. Solar and lunar eclipses therefore tend to occur two weeks apart. However, a lunar eclipse need not be preceded by a solar one, or vice versa because the Moon may not be close enough to a node at the preceding syzygy to eclipse the Sun.

If you have never seen a lunar eclipse, there are a couple of things that you might like to know before you see one. Curiously, the Moon doesn't appear to move into the Earth's shadow in the direction that it should. Although we know perfectly well that it is the Moon that is moving towards the shadow, during the eclipse it appears the other way around. The illusion is brought about because the Moon's actual velocity (it is moving from west to east) towards the shadow is a great deal less than the rate at which the Earth is spinning (also from west to east). The Earth's rotation makes it

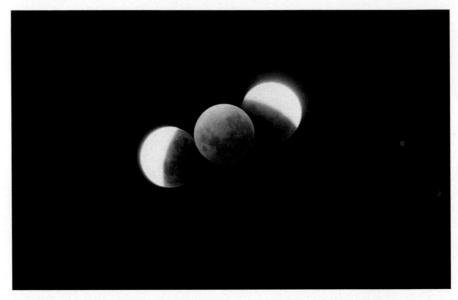

Figure 10.8 Lunar eclipse. A multiple exposure of a partial eclipse of the Moon. Notice how red the Moon appears when in the Earth's shadow. (*Photo* Pekka Parviainen)

appear as if the Moon is moving across the sky from east to west (something that you notice on any night when the Moon is visible). Since the Moon enters the Earth's shadow from the west, our first view of it is to the east of the Moon. During the eclipse, which in all its phases may last a couple of hours or more, the Moon's apparent westerly drift is considerable and this reinforces the impression that the shadow is moving towards the Moon rather than vice versa. You may be struck by how palpably spherical the Moon appears once it is within the Earth's shadow.

The edge of the shadow cast by the Earth on the Moon is circular. Aristotle used this as one of the proofs that the Earth is spherical. He argued, correctly, that only spherical objects always cast circular shadows. If the Earth were a flat disc it would sometimes cast an elliptical shadow.

Unless you know that an eclipse is in progress, you may not notice a penumbral eclipse because there is little change in the Moon's brightness as it passes through the penumbra. Even when totally eclipsed, the Moon is usually visible. Its colour characteristically changes to a dull coppery red because it is illuminated by sunlight refracted into the Earth's shadow by the atmosphere. This light is reddened because a large proportion of its blue component has been scattered out of it in passing through the Earth's atmosphere, the same cause as reddened sunsets. On some occasions, when the atmosphere is heavily laden with volcanic dust, so much of the light passing through the atmosphere is scattered or absorbed that little or no light is refracted into the penumbra, and the Moon vanishes from sight during totality.

The following scale was devised by André Danjon, a French astronomer, to assess the brightness of lunar eclipses. The descriptions apply to the Moon's appearance at mid totality.

Danjon scale of eclipse darkness

0	The darkest eclipse: Moon very dark and nearly invisible at mid totality
1	Moon dark grey or brownish, few details can be made out
2	Moon is dark red or rust red, with darker central areas, outer regions brighter
3	Moon is brick red, frequently its edge is yellowish
4	Moon is coppery red or orange, very bright sometimes with a bluish or greenish edge

Chapter 11 | PLANETS

> Among the many self-luminous moving suns, erroneously
> called fixed stars, which constitute our cosmical island, our
> own Sun is the only one known by direct observation to be a
> central body in its relations to spherical agglomerations of
> matter directly depending upon and revolving around it, either
> in the form of planets, comets or äerolite asteroids.
>
> Alexander von Humbolt, *Cosmos*, Vol. 1, 1845; reprinted John
> Hopkins University Press, 1997, p. 89

11.1 The Solar System

For most people, mention of the Solar System probably conjures up that
well-known image of nine concentric circles, each representing the orbit of
a planet, all more-or-less centred on the Sun. In fact, this image is presented
from a vantage-point that no human has ever occupied, somewhere far
outside the Solar System. From such a distance, the only thing that you
would see of the Solar System is the Sun, reduced to an extremely bright
point of light. Without a telescope you would be unable to see Jupiter, the
largest planet, let alone any of the other planets. The fact is, by any objective measure, such as mass, diameter or brightness, planets are insignificant
motes compared with stars. Yet, what would the Solar System be without
planets? Planets loom large in our view of the cosmos for several reasons,
not the least of which is that we happen to inhabit one of them.

Planetologists, scientists who study planets, have established that planets
are a by-product of star formation, though not every star is necessarily accompanied by planets. A star is the result of the collapse of a huge cloud of gas due
to gravitational attraction between its parts. The Solar System itself began
forming some 4.5 billion years ago from one such cloud of gas, composed of
three parts hydrogen to one part helium, together with traces other substances in the form of ice and dust. The lion's share of the cloud ended up in
the Sun. Most of the remaining material was blasted back into interstellar
space when the Sun's nuclear furnace was ignited. The little that was not
lost was shared out among the various objects that now orbit the Sun.
Planetologists believe that the Solar System reached its present state remark-
ably rapidly, possibly within 50 million years from the start of the process.
This represents a mere hundredth part of the present age of the Solar System.

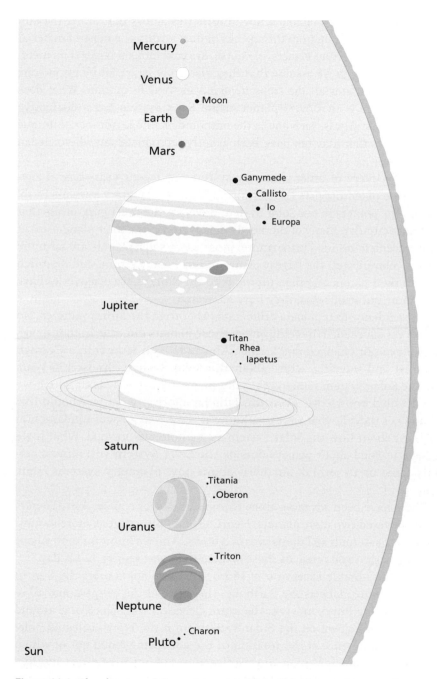

Figure 11.1 The planets and the major moons of the Solar System drawn to the same scale. On this scale the Sun would have a diameter of approximately 33 cm.

Even the tiniest meteoroid a few millimetres across that orbits the Sun was laboriously built up from tiny specks of dust within the mass of material swirling around the Sun. Planets, of course, are enormously larger than meteoroids and, although we assume that they also grew larger bit by bit, no one yet fully understands all the steps involved in their formation. What does seem to be case is that every planet in the Solar System has a distinctive history. In other words, each one is the outcome of a huge number of unique events, events that may not have been exactly reproduced anywhere else in the universe.

The discovery of other planetary systems in recent years has, if anything, made planetologists less certain about how planetary systems form. Whereas the planets in our Solar System have, for the most part, orbits that are almost circular, many of the newly discovered extra-solar planets are in highly eccentric orbits. Furthermore, most extra-solar planets are far more massive than Jupiter, the largest planet in our Solar System, and lie much closer to their parent star than Jupiter is to the Sun. These discoveries have raised more questions than they have answered.

The evidence that planets orbit stars other than the Sun is indirect. No one has yet succeeded in seeing any of these planets directly. In that sense, things have not changed much since Alexander von Humbolt, a German naturalist and traveller, wrote about the Solar System almost 150 years ago. The Solar System remains the only planetary system about which we have detailed knowledge. Even this falls far short of what we should like to know, or indeed what we must know if we are to answer fundamental questions about how the Solar System as a whole was formed. What little we have learned about planets outside the Solar System will almost certainly force us to rethink our ideas about how planetary systems come about.

There have been advances since Humbolt's day, of course. Astronomers have discovered two more planets, Neptune and Pluto, dozens more moons, hundreds of asteroids and thousands of comets. And we know far more about the composition and origin of these objects than was known in his day.

With the notable exception of Pluto, all the planets orbit the Sun in almost the same plane as the Earth, i.e. the plane of the ecliptic; and all of them, including Pluto, move in the same direction: anti-clockwise around the Sun looking down on the Solar System from the North celestial pole. This is a consequence of the rotation of the collapsing cloud out of which these bodies formed. Motion in this direction is known as direct motion. Most asteroids orbit between Mars and Jupiter, and are thought to be the remains of bodies formed during the early history of the Solar System. Some asteroids have orbits that cross the Earth's orbit. The orbital planes of many comets and meteoroids are steeply inclined to the ecliptic.

The four planets closest to the Sun – Mercury, Venus, Earth and Mars – are composed principally of iron and silicon. They all have a solid surface, and are known collectively as terrestrial planets. The two largest planets, Jupiter and Saturn, have a core of ice and rock surrounded by a deep gassy envelope composed mainly of hydrogen. Uranus and Neptune are composed mostly of ice and rock, and contain little gas. Pluto is also composed of ice and rock, though planetary scientists are increasingly of the view that Pluto should not be classed as a planet, but as a huge comet.

The Solar System consists of far more than the Sun and planets. There are, in fact, 25 bodies in the Solar system that have a diameter greater than 1000 km; 26 if we include Ceres, the largest asteroid, which falls just short of this figure by 50 km (its diameter is 950 km). Nine of these are planets, and the remainder, except for Ceres, are moons. Two of these moons – Ganymede, which orbits Jupiter, and Titan, which orbits Saturn – are slightly larger than Mercury and far larger than Pluto. In addition, there are at least some 45 further moons (new ones are discovered from time to time orbiting the larger planets), several thousand asteroids (also known as minor planets), several million comets, and lots of dust (the source of most of the meteors that can be seen every night streaking across the sky.)

The information in the first table will give you an idea of the scale of the Solar System.

Planetary data

Body	Average distance from the Sun (10^6 km)	Average diameter (km)	Sidereal period	Synodic period	Number of natural satellites (2001)
Sun		1.4×10^6			
Mercury	58	3476	88 days	116 days	None
Venus	108	12400	224 days	584 days	None
Earth	150	12600	365 days	N/A	1
Mars	228	6800	687 days	780 days	2
Jupiter	778	142800	11.9 years	399 days	17
Saturn	1428	119400	29.5 years	378 days	28
Uranus	2870	47600	84.0 years	370 days	21
Neptune	4499	48400	164.8 years	368 days	8
Pluto	5900	2200	250.0 years	367 days	1

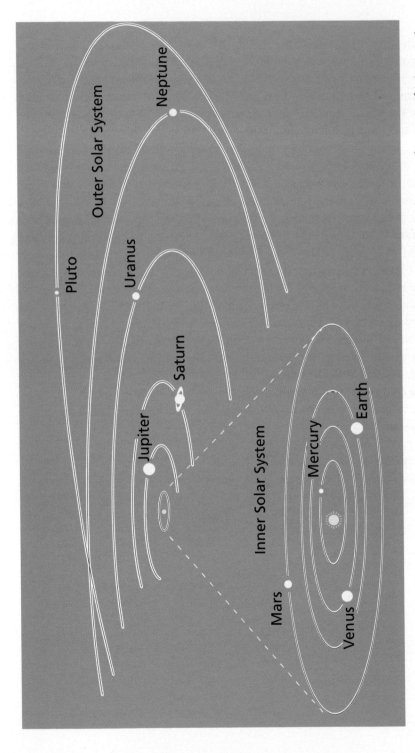

Figure 11.2 The Solar System. Pluto is, on average about 40 times further from the Sun than Earth. The diameter of Jupiter's orbit is approximately 5 times greater than that of the Earth. This makes it difficult to show all the orbits clearly in a single diagram that fits on the page of a book. In this diagram, the orbits of the inner planets are shown separately from those of the outer planets.

Large distances are difficult to grasp. We can conceive them because we can assign numbers to them. But they defy imagination. How can we imagine a million kilometres? You will find that the scale of the Solar System can be more readily visualised by relating all distances and dimensions to the Earth. If the Earth is represented by a sphere 1 cm in diameter (about the size of a small marble) then the Sun becomes a sphere 109 cm across (i.e. 109 Earth diameters), 118 m away (the distance from the Earth to the Sun is 11 800 times the Earth's diameter.)

The relative dimensions of the Solar System are given in the next table. As you look through the table, bear in mind that all dimensions are in terms of the Earth's diameter. Values have been rounded up or down, since in this situation precision gets in the way of the broader picture. To construct a model of the Solar System based on this scale you have only to multiply each value in the table by 1cm. For example, Jupiter will be a small sphere 11.2 cm in diameter at a distance of 61 000 cm (610 m) from the Sun. The most distant planet, Pluto, would be a sphere 2 mm across, 4.6 km away. Large as the Solar System is, the greatest surprise is how far away the stars are. On this scale, the nearest stars – those of the Centauri system – are approximately 32 000 kilometres away, a distance equal to three quarters of the Earth's *actual* equatorial circumference.

Relative dimensions of the Solar System

Body	Average distance to the Sun (in Earth diameters)	Average diameter (in Earth diameters)	Time for one orbit (Earth = 1)	Orbits completed (Earth = 1)
Sun	–	109	–	–
Mercury	4 600	0.4	0.24	4.1
Venus	8 500	1.0	0.6	1.6
Earth	11 800	1.0	1.0	1.0
Mars	17 900	0.5	1.9	0.5
Jupiter	61 000	11.2	11.9	0.08
Saturn	112 000	9.4	29.5	0.03
Uranus	225 000	4.1	84.0	0.01
Neptune	352 000	3.8	164.8	0.006
Pluto	460 000	0.2	248.5	0.004
Centauri star system	3 000 000 000	N/A	N/A	N/A

11.2 How to tell a planet from a star

Planets can disappoint. Unless you know what you are looking for, to the naked eye a planet is just another anonymous bright dot in the heavens, sometimes considerably brighter than other stars, but a dot none the less. All in all, these blazing dots make a most unplanetary impression. Compared with the only planet that we know intimately, our dim, dun-coloured Earth, they seem far too bright, as if they were on fire. The Earth, of course, would look like this seen from far off. The ancients didn't know this, which is why they lumped planets with stars, and never considered that they might be worlds. Even through a powerful telescope, a planet is a fuzzy, mottled disc, nothing like those eye-catching photographs taken by space-craft as they fly past a planet, or by the Hubble Space Telescope.

We can be certain that people have always known about planets, though they have not always known what they were. Planets are sufficiently unlike stars that if you spend most of your time under an open sky, as people used to, sooner or later you can't help but pick them out among other bright points of light that are strewn across the night sky. What distinguishes them from stars is that they don't stay put. Over a period of a few days, a planet visibly alters its position relative to the fixed stars. In fact, that is how they came to be named planets.

The word 'planet' comes from a Greek word meaning wanderer. Astronomers in ancient Greece believed that planets were really stars, but called them planets, or 'wandering' stars, because they appear to move among the fixed stars. As far as these early astronomers were concerned, the only difference between a star and a planet was that a planet moved around the Earth in the space between the Earth and sphere of the fixed stars, whereas stars stayed put. In every other respect, planets were considered to be similar to the fixed stars: all celestial bodies were made of the same sub-stance, a fifth element, known as 'ether', and, except for the Sun, which was obviously self-luminous, were visible because they reflected sunlight. The Sun and the Moon were also considered to be planets, since they too move around the sky.

Ancient astronomers were wrong about almost everything to do with the nature and motion of planets and stars, though it took quite a while for their views to disappear from circulation, even after the truth was known. The truth, of course, is that the Earth is a planet, and that it orbits the Sun together with several other planets. This is not obvious when we look at the night sky because the sky seems to revolve around the Earth. However, as Copernicus realised, a satisfactory explanation of the movements of the 'wandering stars' requires that Earth be displaced from its position at the centre of the cosmos, and join them in orbit about the Sun.

It didn't take long for astronomers to realise that, in joining these wandering stars, the Earth conferred on them the status of worlds. Planets could no longer be considered to be stars. After all, it was argued, if the Earth orbits the Sun like any other planet, then it is more than likely that planets share something of Earth's nature: they also must be worlds, possibly inhabited. In 1609, seventy years after Copernicus published his ideas, Galileo turned his telescope on the heavens and saw that planets are not mere featureless dots. Even a low-power telescope will show faint bands on the surface of Jupiter, the ring around Saturn and the ice caps of Mars.

But even the best ground-based telescopes are unable to form a really clear image of a planet because the Earth's atmosphere is never still. Ceaseless currents of air continuously alter the passage of light through the atmosphere, making it all but impossible to form a clear image of anything that lies beyond. It's only since planets were photographed from spacecraft, within the last 25 years, that we have been able to see their surfaces clearly enough to see what they are really like.

We now accept that planets hold up a mirror to the Earth: like them the Earth too moves through the void, held in its orbit by a mutual gravitational embrace with the Sun. The sight of a planet is an occasion to reflect on Earth's real nature. Like them, Earth is a fragment of matter, isolated and vulnerable in the immensity of space, endlessly circling the Sun. Unlike them, it is the only planet known to support life.

We can see six of the planets of the Solar System with the naked eye. Their apparent brightness varies depending on where they are relative to the Earth and Sun. Mercury, Venus, Mars, Jupiter and Saturn can all shine more brightly than the brightest star for several weeks at a stretch, and consequently have been known since the earliest times. The sixth, Uranus, can attain magnitude +5 (see section 12.3 for more details about magnitude) when at opposition, the point when it comes closest to the Earth, and so hovers at the limit of what is visible to the naked eye for weeks at a time. Nevertheless, it remained unknown until its accidental discovery by telescope in 1781 by Sir William Herschel, one of the greatest astronomers of all time. Interestingly, after its discovery, searches of astronomical records revealed that other astronomers had seen Uranus on a number of occasions in earlier times, but that none of them had recognised it for what it was. For example, John Flamsteed, the first Astronomer Royal, had made a note of the planet during a survey of the sky in 1690, but had not followed up the observation.

Even Herschel at first thought that what he had found was a comet. A distant comet and a planet can look similar when seen through a telescope because both appear as fuzzy discs, and Herschel had no reason to believe that he had discovered a new planet. He hadn't been looking for one. In his day the only new objects that astronomers expected to discover orbiting the Sun were comets. They knew nothing of asteroids (the first asteroid was discovered in 1801). But calculations based on measurements of its motion revealed that the orbit of Herschel's 'comet' was almost circular, like that of other planets. A comet's orbit is very different: an elongated ellipse that takes it close to the Sun at one extreme, and far away at the other.

Now that its orbit is known, given clear skies and a favourable opposition, i.e. when Uranus and Earth are both on the same side of the Sun, and as close as possible, a determined observer should be able to see Uranus with the naked eye. To see Neptune (discovered in 1846) you will, at very least, have to use binoculars, and Pluto (discovered in 1930) is visible only with a powerful telescope. Astronomers now agree that there are no planets orbiting the Sun beyond Pluto.

This is a good place to mention that it is possible to see some of the other larger permanent members of the Solar System with the naked eye given very good seeing conditions. Jupiter has four moons that are easily seen with binoculars as tiny points of light. They are, listed in increasing distance from Jupiter: Io, Europa, Ganymede and Callisto. They all move rapidly, and if you keep an eye on them over a few days you'll notice that they alter position from one day to the next. Records show that Callisto and Ganymede, the two largest moons of Jupiter, have been seen with the naked eye, though not on the same occasion. Vesta, the brightest, though not the largest, asteroid, can attain magnitude 5.1, and has sometimes been seen with the naked eye. The largest asteroid, Ceres, is about twice as large as Vesta, and can attain magnitude 6.7, which makes it too dim to be seen without binoculars. Clearly, Vesta and Jupiter's moons are not objects that you are likely to see by chance with the naked eye. They can be seen only by a well-prepared, determined observer. If you wish to try your hand at spotting these faint objects, their position is given in handbooks published by astronomical associations, and sometimes in astronomy magazines. Alternatively, use a computer simulation of the sky or the internet.

If you are unfamiliar with the night sky, or if you don't follow what is going on in it, you will find it difficult to tell a planet from a star with a casual glance. Nevertheless, there are several tell-tale signs that will help you distinguish between planets and stars with the naked eye. These are as follows.

- A planet generally has a brighter, steadier image than a star. Stars tend to twinkle, whereas planets don't, unless they are close to the horizon.
- The apparent brightness of a planet can change very noticeably in the course of a few weeks.
- When it attains maximum brightness, a planet will be the first celestial object to become visible in the sky after sunset, and it will be the last to vanish at dawn.
- The position of a planet, seen against the background of stars, can change noticeably from one night to the next.
- A planet will always be seen on, or very near, the plane of the ecliptic.
- A planet can be seen with the naked eye, even when it is within a few degrees of the horizon.

11.3 Inferior and superior planets

There are several ways in which planets can be classified or grouped. A planetologist, interested in their origins and composition, would class them as either terrestrial, or giants of gas or ice. To an observer on Earth, particularly someone who merely wants to know where to look in order to see these planets, there is a more fruitful way of grouping them. This is in terms of where their orbits lie with respect to Earth. In this scheme, Venus and Mercury are said to be inferior planets because their orbits lie between the Sun and the Earth. The remainder, Mars, Jupiter, Saturn, Uranus, Neptune and Pluto, are known as superior planets because their orbits lie beyond Earth's orbit.

When describing the motion of either inferior or superior planets around the Sun, certain of their positions with respect to the Earth are of particular importance because these can be unambiguously defined, and so provide valuable reference points. They are shown in figures 11.3 and 11.4. Because both the planet and Earth are in motion, to understand what you will see when you look at a planet from the Earth you have to consider the effect of two orbital motions together: that of the planet and that of the Earth.

The time taken for a planet to complete an orbit of the Sun with respect to the fixed stars is its sidereal period. The time it takes to return to the same place in the sky as seen from the Earth is its synodic period. The Earth's sidereal period is, of course, 365.25 days. If you want to know when you are likely to see a planet then its synodic period is of greater significance than its sidereal period because this determines when it will next return to the same place in the sky seen from the Earth.

11.4 Where to look for an inferior planet

There are three factors that determine where in the sky we should look to see an inferior planet. The first is that an inferior planet orbits much closer to the Sun than the Earth does. Hence it is always seen near to the Sun, in either the evening or the morning sky, and never in the middle of the night, except at very high latitudes.

The second factor that affects the visibility of an inferior planet is the inclination of the ecliptic to the horizon. Since all planets orbit the Sun more or less in the plane of the ecliptic, the visibility of an inferior planet depends to some degree on the time of year at which it reaches maximum elongation. Just as in the case of the young Moon, the best views are to be had when the ecliptic is steeply inclined to the horizon. If you live north of the equator, both Venus and Mercury are highest in the sky if they are evening stars in spring, or morning stars in autumn. The situation is reversed south of the equator.

Finally, there is the speed with which the inferior planet orbits the Sun. An inferior planet moves faster than the Earth and so completes its orbit in under a year. All planets move around the Sun in the same direction, and so an inferior planet has to complete more than one orbit in order to return to the same position in the sky as seen from the Earth. The time it takes to do this is, of course, its synodic period, which is 584 days for Venus, and 116 days for Mercury.

We can now work out in more detail how the motion of an inferior planet around the Sun determines where in the sky you are likely to see it. Let's begin at the moment when the planet is hidden from Earth on the other side of the Sun. Astronomers call this alignment between Earth, Sun and an inferior planet 'superior conjunction'. Since it orbits in the same direction as the Earth, an inferior planet makes its first appearance in the evening sky at sunset. Thereafter, over a period of weeks, it gradually moves out of the Sun's glare and into the twilight of the evening sky. On each successive day the apparent distance between the planet and the Sun increases, though the rate at which this happens is not constant. At the same time the planet becomes increasingly prominent because we see it against the darker sky of late evening. After several weeks, it begins moving back towards the Sun, eventually setting with the Sun. The turning point occurs when the planet reaches its greatest eastern elongation. Geometrically, greatest elongation corresponds to the point at which the line of sight from the Earth to the planet forms a tangent to the planet's orbit (see the figure 11.3).

The planet takes longer to reach eastern elongation from superior con-

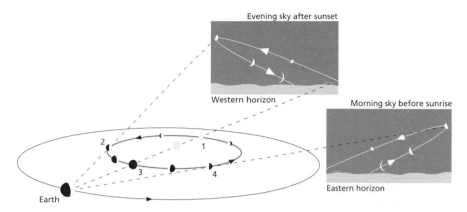

Figure 11.3 The apparent path of an inferior planet seen from the Earth's surface is determined by the relative positions of the Earth and the inferior planet. When the planet is on the far side of the Sun it is at superior conjunction (1). When it is between the Earth and the Sun it is at inferior conjunction (3). Greatest eastern elongation, when the inferior planet appears to be as far east from the Sun as it can get, occurs at 2. Greatest western elongation is at 4.

junction than it does to return to inferior conjunction from eastern elongation. It's not difficult to see why. In effect we see the planet's orbit edge-on. Hence, over a period of many months, we see it move back and forth like a pendulum swinging across our line of sight to the Sun. Since it has to travel further along its orbit to get from superior conjunction to greatest eastern elongation than it does to return from this elongation to the Sun, it appears to move away from the Sun slowly and return to it quickly. Venus takes approximately seven to eight months to reach its greatest eastern elongation from superior conjunction, and only some two and a half months to return to inferior conjunction. The comparable figures for Mercury are five weeks and three weeks respectively. After inferior conjunction, the planet re-emerges from the Sun's glare into the morning sky, reaches its greatest western elongation relatively quickly, and finally returns to superior conjunction more slowly.

Seen through a telescope, both Venus and Mercury go through a cycle of phases, like those of the Moon. Both planets show their full face when at or near superior conjunction. However, they are then at their greatest distance from the Earth, and so are not very bright. As they approach eastern elongation, they become gibbous, and grow brighter as their apparent size increases. They reach quarter phase at eastern elongation, and then go on to become crescents. Venus reaches its maximum brightness shortly after it has passed its greatest eastern elongation, and again before it reaches its greatest western elongation.

Visibility of an inferior planet

Planet's position relative to the Sun and Earth	Planet's position seen from the Earth
Superior conjunction	Not visible
Heliacal rising	Becomes visible above western horizon (as an evening star)
Greatest eastern elongation	Attains greatest distance east of the Sun and is therefore visible for the greatest time after sunset
Heliacal setting	Sets in the west with the Sun
Inferior conjunction	Not visible
Heliacal rising	Becomes visible above the eastern horizon (as a morning star)
Greatest western elongation	Attains greatest distance west of the Sun and is therefore visible for the greatest time before sunrise
Heliacal setting	Rises in the east with the Sun
Superior conjunction	Not visible

11.5 Mercury

Mercury's orbit is highly eccentric, which means that there is a large difference between its maximum and minimum distances from the Sun. Consequently, its greatest elongation varies from 16° to 28°. However, even at its maximum elongation, it is a difficult object to see because it always lies within the twilight arch. By the time the Sun has dropped far enough below the horizon for the twilight arch to darken, Mercury will itself be within a few degrees of the horizon, where it is likely to be lost from sight because of horizon haze. This is probably the principal reason why it is the least seen of the visible planets. In the northern hemisphere, the best chance of seeing it always occurs when it is an evening star at the time of the spring equinox, or when it is a morning star at the autumnal equinox. You will find that a pair of binoculars is a great help in picking it out. Once you have located it, you should have no difficulty seeing it with the naked eye.

11.6 Venus

The orbit of Venus is almost circular, so its greatest eastern and western elongations are practically the same: 46°. At these times it is seen against a dark sky, and this, coupled with the fact that it is completely covered in cloud, which makes it highly reflective (it reflects 70% of the sunlight illuminating it), makes it the brightest object in the sky after the Sun. Indeed, at greatest elongation and on a moonless night, it is bright enough to cast a shadow. These shadows are difficult enough to see even when there is no Moon and the sky is clear. Artificial lighting makes them impossible to see. However, they were regularly seen in the days before our towns and cities were brightly illuminated with electric light, and a moonless night really was dark. Gilbert White made a note in his diary of seeing them in 1782.

> Feb. 8 Venus shadows very strongly, showing the bars of the window on the floors and walls. Feb. 9 Venus shed again her silvery light on the walls of my chamber and shadows very strongly.
>
> Gilbert White

At its brightest, Venus is visible during daylight if the sky is very clear, and if you know exactly where to look. It is seldom seen by accident, so you need to know its elongation and its angular height above your local horizon on the day that you intend to look for it. The best source of this information is a computer simulation of the sky. Choose a date when the planet's elongation is 40° or more: Venus will then be bright and sufficiently far from the Sun that you won't look at the Sun by accident. Using the values for elongation and elevation obtained from the computer simulation, determine the approximate position of Venus relative to the Sun and horizon using your outstretched hands to measure angles (see Appendix). Then search that stretch of the sky with binoculars. Once you have found Venus with the binoculars, something that it is not difficult to do, you should be able to see it with the naked eye as a faint point of light, so long as you have good eyesight. You will find that it is very easy to lose sight of the planet if your eyes wander from it.

You should be extremely careful when using binoculars to look at the sky when the Sun is above the horizon. Venus is never far from the Sun, and, unless you take precautions, you may inadvertently look at the Sun while scanning the sky for the planet. Be warned: if you look at the Sun through binoculars, even briefly, you will almost certainly damage your eyesight permanently. You can prevent this happening by standing well inside the shadow of a building or tree so that you can't see the Sun, but can see the portion of sky that contains Venus. Use the building to help locate Venus: line up the edge of the building with the approximate elongation of the planet, minus a few degrees.

11.7 Where to look for a superior planet

A superior planet moves around the Sun much more slowly than the Earth does. Consequently, a superior planet is overtaken by Earth several times in the time it takes the superior planet to complete an orbit. As a result, a superior planet will not have moved very far along its orbit by the time the Earth catches up with it again, which makes its synodic period similar to the Earth's sidereal period of 365 days. The exception to this is Mars because it moves significantly faster than the other superior planets.

Except for Mars, therefore, all superior planets are visible in the night sky for several weeks each year. Unfortunately, they are not always favourably placed for observation because they too are continuously moving along their orbit and so don't return to exactly the same position in the sky each year. The best sightings are had when they reach opposition during winter because the ecliptic is then high in the sky at midnight and so the planet will be well above the horizon. Superior planets can be seen in the midnight sky because their orbits lie outside the Earth's orbit.

It's relatively simple to work out, in general terms, the apparent path of a superior planet across the sky. Suppose we begin by considering the situation at conjunction. The planet is then not visible to us since it is on the other side of the Sun from the Earth. Although the planet orbits in the same direction as the Earth, it moves more slowly, so that it first becomes visible from Earth just before dawn on the eastern horizon. Over a period of several weeks, as the Earth gradually catches up with the planet, the planet rises ever earlier ahead of the Sun. During this period it appears to drift eastward through the starfield against which it is seen. This eastward drift is known as direct motion.

The eastward drift is reversed after the planet reaches western quadrature. At western quadrature the planet rises on the eastern horizon at midnight. At this point the Earth is moving directly towards the superior planet and for several days the planet's position against the fixed stars changes slowly. This is known as a stationary point. As the planet moves towards opposition, the Earth is overtaking the superior planet, which consequently stops drifting eastward and now appears to drift westward through the starfield. Motion in this direction is known as retrograde motion. Retrograde motion was one of the things that Copernicus set out to explain in his heliocentric theory, successfully as it turned out.

At opposition, the superior planet rises on the eastern horizon at sunset, and remains visible throughout the night. It also attains its greatest brightness because it is closest to Earth and the whole of its illuminated disc faces

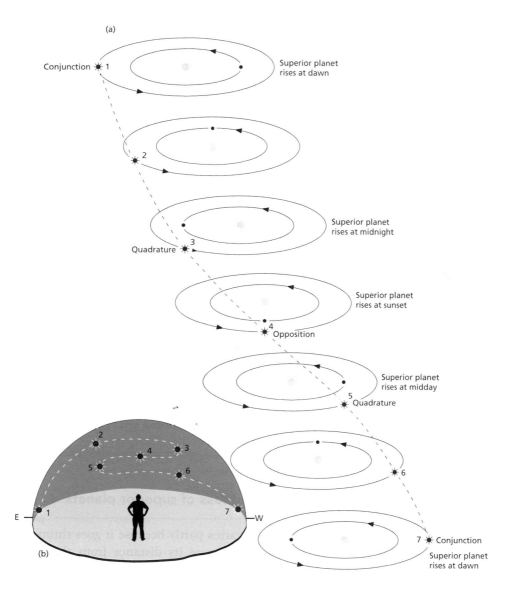

Figure 11.4 (a) shows the relative positions of the Earth and superior planet as they orbit the Sun. The apparent motion of a superior planet across the sky is due to the relative positions of the Earth, Sun and the superior planet.

(b) shows where you will see the superior planet in the night sky from the ground over several weeks.

Chapter 12 | STARS

It's lovely to live on a raft. We had the sky up there, all speckled with stars, and we used to lay on our backs and look up at them, and discuss about whether they was made or only just happened. Jim he allowed they was made, but I allowed they happened; I judged it would have took too long to MAKE so many. Jim said the Moon could a LAID them; well, that looked kind of reasonable, so I didn't say nothing against it, because I've seen a frog lay most as many, so of course it could be done. We used to watch the stars that fell, too, and see them streak down. Jim allowed they'd got spoiled and was hove out of the nest.

Mark Twain, *The Adventures Of Huckleberry Finn*,
Penguin, 1994, ch.12, p. 120

12.1 Light without form

The price we pay for city life is blank urban night skies, rendered almost starless by our addiction to light. Electric light is, without doubt, a 'good thing'. But, like all good things, you can have too much of it. Many of our cities are so brightly lit that our eyes are perpetually dazzled, and we are unable to see any but the very brightest stars and planets. If you live in a city, you may have forgotten what darkness is really like. Worse, the glow of one city often joins up with that of others so that, even from the countryside between them, few stars are visible. We are gradually being denied sight of the stars, a birthright prized by our ancestors. It's a sad thought that future generations may know the grandeur of the heavens only from hearsay.

Before the modern era of astronomy, which began with Copernicus in the sixteenth century, it was assumed that all stars were equidistant from Earth and that they marked the outer edge of the Universe. Lacking the apparatus and scientific knowledge of the modern astronomer, the ancients could only speculate about the nature of these distant points of light. Plato and his followers believed that stars were composed of fire, one of the four elements of which they believed the world was made. But most people bowed to the authority of Aristotle, who argued that the heavens, stars and planets were made of aether, an incorruptible fifth element not present on Earth.

Astronomers who came after Copernicus were at first less concerned with the nature of stars than they were with mapping their positions accurately. The purpose of these maps was primarily to aid navigation, and hence facilitate trade. It was generally accepted that stars were like the Sun, though this wasn't much of an advance on Plato or Aristotle, since the nature of even the Sun remained a complete mystery until the latter years of the nineteenth century. The very remoteness of even the nearest stars meant that astronomers knew little more than Huckleberry Finn about what stars were.

Early in the nineteenth century, Auguste Compte, an influential French philosopher, rashly claimed that: 'We conceive the possibility of determining . . . [the stars'] . . . distances, their magnitudes, and their movements, but we can never by any means investigate their chemical composition'. In other words, astronomers would never discover what stars really are. Compte felt he was on sure ground here because he knew that it is not possible to see stars for the objects they are. Even with a powerful telescope all stars stubbornly remain points of light. In fact, we could call their light 'light without form' to emphasise this fact. We associate light with form, and we expect to see things when they emit or reflect light. Our eyes focus this light and form images. We see objects, not merely light. Seeing a thing allows us to identify it as a house, or a tree or a mountain. Starlight confounds these expectations since stars can't be imaged even at the highest magnifications of which the most powerful modern telescopes are capable. No wonder Compte was so confident in his scepticism. That he did not foresee how astronomers would discover that it is possible to tease the secrets of the stars from a few photons collected by a telescope is excusable.

The key to unlocking the secrets of these distant objects was found to be in their light: the spectrum of light from a star contains clues to its composition, temperature, motion, and much else. The techniques and methods by which this information could be gleaned from light were being developed even as Compte wrote those fateful words. The first star to which spectral analysis was successfully applied was the Sun. In 1859 Gustav Kirchhoff, a distinguished German physicist, discovered that there was sodium on the Sun by looking at the spectrum of sunlight. By 1891 a further 35 elements in the Sun had been identified. The application of spectral analysis to starlight began in 1864, when an English astronomer, William Huggins, examined some nebulae and concluded that they are composed of glowing gases, and are not groups of distant stars.

We now know that every star is a huge sphere of gas – mainly hydrogen with a significant amount of helium – heated to incandescence (white heat) by a nuclear reaction at its core. The reaction converts hydrogen into helium at unimaginably high temperatures, while gravity keeps the whole from exploding into the surrounding void.

Stars differ from one another in size and brightness. The largest have diameters that are tens, if not hundreds, of times that of the Sun, and are also the brightest. The smallest are probably not much larger than Jupiter, a tenth of the Sun's diameter, and are so dim that they are almost impossible to spot even with a powerful telescope.

The surface of a typical star is at a temperature of several thousand degrees, far cooler than its core, and emits mainly light and infrared radiation, and some ultraviolet. When you look up at the night sky, it is difficult to believe that each of those tiny, sparkling points of light is fuelled by an unimaginably powerful nuclear furnace – until you realise that the Sun is also a star, the only star close enough for its surface to be visible to us.

All stars begin in the same way: a vast cloud of gas and dust is drawn together by gravitational attraction over hundreds of thousands of years. As the gas is squeezed into an ever-smaller volume it heats up, until the temperature at the centre of the cloud reaches a point where atoms of hydrogen can fuse together to form atoms of helium. This energy of this reaction checks the gravitational collapse, at least until all the hydrogen in the core has been converted into helium. The reaction is fiercest in the largest stars, which is why they are short lived, lasting only a few million years. The stars with the longest lifespan are the smallest; calculations based on the rate at which nuclear fusion occurs at the comparatively low temperature found in the core of a small star suggest that the smallest stars can last trillions of years before they run out of hydrogen.

When the hydrogen in the core is finally exhausted, the nuclear fire is extinguished, and gravity can once more squeeze the star into a smaller volume, and in the process increase the temperature at its core. In all but the smallest stars, the gravitational squeeze re-ignites a nuclear reaction, more powerful than the first, and the star expands to several times its original volume, becoming a red giant. But no star contains enough fuel to keep the nuclear fires burning forever, and sooner or later all red giants die spectacularly when gravity finally crushes them into a fraction of their original volume. A Sun-like star will, some 5 billion years hence, become a white dwarf, an enormously hot, inert body, about the size of Earth. The very largest stars often collapse completely and become black holes, mere points in space exerting huge gravitational force.

Almost all the points of light that you can see in the night sky are individual stars. However, distant groups of stars can also look like individual stars to the naked eye because their fainter parts are not bright enough to be visible. Binoculars can reveal something of the shape and size of the brightest of these objects. The Andromeda Galaxy, which under good seeing conditions appears to the naked eye as faint, fuzzy star, becomes a diffuse, elongated patch of light when seen through binoculars, and has an apparent size similar to that of a full Moon. See the list of non-stellar objects on page 280 for further examples of star-like objects that are really something else.

Since astronomers in ancient Greece assumed that all stars were equally distant from the Earth, they took the brightness of a star to be an indication of its size because, to the naked eye, the brightest stars appear to be considerably larger than the dimmest ones. The system of stellar magnitude used to classify stars, which was originally devised in the second century B.C. by Hipparchus, appears to have been based on just this assumption. In fact the assumption is correct: the larger a star, the greater its intrinsic brightness. But since stars are not all equidistant from the Earth, their apparent brightness can't be taken as an indication of their intrinsic brightness: a star that appears bright is not necessarily very large – it may be closer to us than other stars.

The advent of the telescope put an end to the idea that it was possible to measure the diameter of a star by direct observation because, even with the largest instrument, it is impossible to see the surface of a star. A planet, when seen through a telescope, shows a definite disc but, even under the best seeing conditions, stars always appear as points of light of differing brightness. A telescope merely makes a star brighter.

What makes a bright star look larger to the eye than a dim one is irradiation. Light is scattered within the retina at the back of the eye, and so light from a bright star stimulates retinal cells that lie outside the immediate area within which the star's image is focused. The same thing occurs in a photographic emulsion, which is why, in photographs of the night sky, bright stars form larger images than dim ones.

12.2 Where are the stars?

It's tempting to believe that when we look at the night sky we are gazing into an infinite abyss. In fact we see only darkness, which is an altogether different thing. Darkness on its own is merely an undefined void, perceived by eyes that have nothing to focus upon. You can experience darkness simply by closing your eyes. What makes the void of the sky tangible are the stars scattered through it, which act as beacons that give us some sense of the vastness we are contemplating.

It has been estimated that 95% of all stars are smaller and cooler than the Sun. Being cool, they are not very bright. Almost all stars that you can see with the naked eye are intrinsically far brighter than the Sun, which makes them visible at great distances. Most of the stars that lie closest to the Sun are not bright enough to be seen without a telescope. Of the 50 stars within 5 parsecs of the Sun, only nine are visible to the naked eye. Among these is Sirius, which appears to be the brightest star in the night sky. However, its apparent brightness is due principally to the fact that it is a near neighbour – a mere 2.7 parsecs away from the Sun. In terms of intrinsic brightness, Sirius ranks 90th among the 100 intrinsically brightest stars visible to the naked eye over the entire celestial sphere.

Among stars with the greatest intrinsic brightness that you can see with the naked eye is Deneb, the brightest star in the constellation of Cygnus. If Sirius and Deneb were both placed side by side, Deneb would appear some 9 times brighter than Sirius, bright enough to cast noticeable shadows at night. In terms of apparent brightness, however, Deneb, which is approximately 500 parsecs from the Sun, about 200 times further away than Sirius, ranks 17th among the brightest stars visible to the naked eye, and appears to be about 3 times less bright than Sirius.

In fact, Deneb is one of the most distant stars that can be seen with the naked eye. You can see individual stars that are more distant, but these appear much fainter than Deneb, and lack a proper name, which would help a casual observer to pick them out from surrounding stars. Instead they are known only by a number in a star catalogue. All distant stars visible to the naked eye share one characteristic: they are intrinsically enormously bright because they are as large as stars can get: they are so-called 'supergiants'.

All the stars that you can see in the night sky are members of the same star system, a spiral galaxy we call the Milky Way. A spiral galaxy is a flattened disc resembling a rotating catherine wheel, and is composed of gas and dust and billions of stars. Most stars are confined to the spiral arms. It used to be thought that the Milky Way was among the largest galaxies in the Universe. It is now known that this is not the case: the Milky Way is of average size having a diameter of approximately 100 000 light years (or 31 000 parsecs).

Now consider how far into this star system we can see with the naked eye. As far as individual stars are concerned, we can't see much further than Deneb, a very near neighbour, just 500 parsecs away. This is a mere $\frac{1}{100}$ the radius of the Milky way. Think how things appear on Earth when visibility is reduced by a fog until it is only $\frac{1}{100}$ the distance to the horizon. If you were on a level plane, where the horizon would be about 6 km away, you would

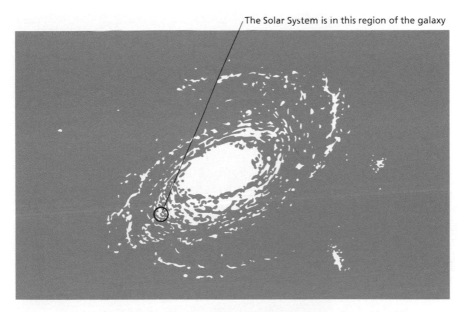

The Solar System is in this region of the galaxy

Figure 12.1 The Milky Way Galaxy. This is a large spiral galaxy consisting of huge amounts of hydrogen and helium gas together with some hundred billion stars, all of which rotates about a centre that may well contain a massive black hole. The Solar System is about $\frac{2}{3}$ from the centre. On this scale it would be about the size of one of the atoms in the paper on which this illustration is printed. (A sheet of paper is about one million atoms thick.)

only be able to see objects within 60 metres of where you were standing. Beyond that, objects would be too faint to be seen, and nearby objects would be seen against a featureless background.

What about the Solar System? Where is it in the Milky Way? The Sun is approximately 27 000 light years (or 8400 parsecs) from the centre of the Galaxy. The Earth's orbital plane about the Sun – the plane of the ecliptic – is inclined at approximately 60° to the plane of the Milky Way, the so-called galactic plane, and is parallel to the Galaxy's radius. We get to see slightly different parts of the galaxy at different times of the year because of the Earth's motion around the Sun. In summer, the Earth lies between the Sun and the centre of the Galaxy whereas, in winter, the Sun lies between the Earth and the galactic centre. In winter, the night sky is filled with bright stars because we are then looking into the spiral arm, known as the Orion arm, which contains the Sun. In summer, the night sky lies in the direction of the centre of the Milky Way. However, this is not directly visible to us because of intervening clouds of interstellar dust. We see instead the Sagittarius arm. The stars in this arm are further from us than those in the

Figure 12.2 The Milky Way.
From our vantage point
within the Milky Way Galaxy
we see it as a faint band. In
this photograph the streaks of
light are bright meteors.
(*Photo* Pekka Parviainen)

Orion arm, so we see fewer individual bright stars. However, the Milky Way looks spectacular in this direction because we are looking towards the centre of the Galaxy with its greater concentration of stars. In spring and autumn, the night sky faces away from the galactic plane, and so these skies contain fewer bright stars than those of summer or winter.

From our vantage point within the Orion arm we see our galaxy as a faint, nebulous band encircling the celestial sphere. Binoculars reveal that this nebulosity is in fact made up of a vast number of individual stars. The orientation of the galactic plane with respect to the horizon changes throughout the night because of the Earth's rotation, and from one night to

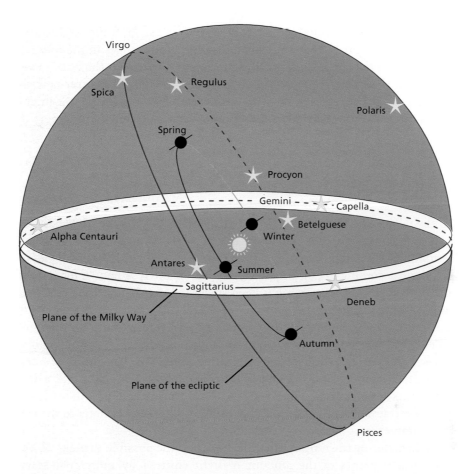

Figure 12.3 The orientation of the Earth's orbital plane within the Milky Way Galaxy. The Earth orbits the Sun in a plane that is inclined at about 60° to the plane of the Galaxy. The only stars shown here are the brightest that lie on the great circles of the ecliptic and galactic plane. The centre of the Milky Way lies in the direction of the constellation of Sagittarius, and the outer edge of the Milky Way lies in the opposite direction, towards the constellation of Gemini.

the next as the Earth orbits the Sun. Since the galactic plane lies at an angle to the ecliptic, you can never see the whole Milky Way from high latitudes because some of it is always hidden from view by the Earth beneath your feet.

The apparent motion of stars from one night to the next is due to the Earth's motion, and is explained in chapter 8. Their immobility with respect to one another is an astronomical illusion. Stars are actually moving at huge speeds relative to one another but, because they are so far from us, it would take several hundred years, i.e. several lifetimes, for changes in the relative positions of even the nearest stars to become noticeable to the naked eye. Motion of these nearby stars is, however, detectable over shorter periods through careful measurement with a telescope. By and large, the night sky of our era is indistinguishable from that seen by Greek astronomers in 200 B.C.

12.3 Star brightness

To the naked eye, the most distinctive feature of a star is its brightness, so it is not surprising that when, in the second century B.C., the Greek astronomer Hipparchus compiled the very first catalogue of stars he classified them according to how bright they appeared. Little is known for certain about his methods, though it has been suggested that he used the fact that stars become visible in stages in the gathering darkness after sunset. Stars that are seen soonest after sunset were classed as stars of the first magnitude. Some while later, with fading twilight, other stars become visible. These would be classed as second magnitude, and so on. The dimmest stars, of course, become visible only when it is completely dark. These were designated sixth magnitude. The use of the word 'magnitude' for brightness is almost certainly based on the fact that brighter stars appear larger to the eye than dimmer ones because of irradiation.

When, during the nineteenth century, it became possible to make objective measurements of the amount of light emitted by individual stars, astronomers noticed that a typical first magnitude star is almost 100 times brighter than a sixth magnitude star. The reason why it doesn't appear to be a hundred times brighter to the naked eye is that the eye's response to changes in brightness is not linear. In mathematical terms, the eye responds arithmetically to a geometrical change in the output of the source. Hence, if the output of a source increases in steps of 2, 4, 8 and 16, the eye perceives this as steps of equal brightness. Increasing the output of a source from 2 to 4 units of light energy is perceived by the eye as the same change in brightness as increasing its output from 8 to 16 units of light energy. In other words, for the eye to notice the same change in brightness in a source of light, the output of the source must be increased by a larger amount when it is bright than when it is dim.

When astronomers got around to measuring rather than estimating the brightness of a star, they could have abandoned Hipparchus' system of stellar magnitude. Instead, they chose to put his system on a firmer footing. To do this, it was proposed that a first magnitude star should be exactly 100 times brighter than a sixth magnitude star. Since this difference in brightness corresponds to five intervals of magnitude, each successive magnitude must be the fifth root of 100 times dimmer than the previous one. The fifth root of 100 is 2.51. Thus a first magnitude star is almost exactly 2.5 times brighter than a second magnitude star, 6.3 (i.e. 2.5×2.5) times brighter than one of the third magnitude and 15.8 (2.5×2.5×2.5) times brighter than one of the fourth magnitude, and so on. Having established a mathematical relationship between magnitude and brightness, the magnitude of any star can be precisely assigned by measuring its brightness. There are, however, two annoying legacies from Hipparchus' system. In the first place the zero point of the scale is arbitrary, and secondly, increasing positive values are used for decreasing brightness. On this scale the four brightest stars actually have negative magnitudes (e.g. Sirius has a magnitude of −1.4). All other stars have positive magnitudes.

Few stars have whole-number magnitudes. This fact is not always immediately apparent when you look at a star atlas. For the sake of simplicity, stars in maps of the night sky are often represented by symbols giving the nearest whole-number magnitude.

The magnitudes in star maps are apparent magnitudes. Apparent magnitude is, as you would expect, a measure of apparent brightness. However, a star may appear bright either because it is close, though intrinsically dim, or far away and intrinsically bright. To calculate intrinsic brightness from apparent brightness it is necessary to know how far the star is from the Sun. This has been measured for many stars. Because astronomers have not entirely abandoned Hipparchus' system, the intrinsic brightness of a star is given in terms of a scale of magnitude known as absolute magnitude.

If you are unfamiliar with the magnitude scale, but you'd like to get some feel for it, go outside on a clear night, and pick out several stars that you consider are equally bright. Check their apparent magnitudes in a star map. How good are you at choosing stars of similar brightness without prior knowledge of their magnitudes? Using a star atlas, choose three stars that are close to one another in the sky and that differ in brightness from one another by one magnitude. Use the star with the intermediate magnitude as a reference point, and compare the difference in brightness between it and the star of one magnitude greater and one magnitude less. Do the intervals of brightness appear to be similar in both cases?

The magnitude of a celestial object is a guide to how likely it is that you will see it. According to convention, the faintest celestial objects that one is

A final point: a telescope makes it possible to see stars that are too faint to be seen by the naked eye because it increases the brightness of a point source. However, it cannot make an extended object visible if it is too faint to be seen by the naked eye because it can't increase its brightness for the reasons given above. You can, however, photograph these faint objects because a photograph can collect light from an object over a long period of time.

12.6 Why do stars twinkle?

A star twinkles because air turbulence affects the path its light takes through the atmosphere. Air turbulence is due to motion between layers of air that are at different temperatures. Friction between layers creates ripples at the boundary between them. In crossing this boundary, starlight becomes concentrated where the layer of colder air is convex. These atmospheric ripples move across your line of sight with the result that the brightness of a star seen through them alters periodically: it twinkles. The rapidity of the twinkling is evidence that the wavelength of the ripples is small (a few centimetres at most).

The technical term for twinkling is scintillation. Scintillation also causes colour-separation because blue light is refracted more than red light. Thus a twinkling star usually goes through rapid periodic colour changes. These changes are much more obvious if you look at a twinkling star through binoculars, while moving them around in small circles. Because of the persistence of vision, the star's image is smeared out into an ellipse made up of a succession of colours separated from one another by dark space. The dark spaces correspond to the period when scintillation prevents starlight from reaching the eye.

Twinkling is most marked in stars that are close to the horizon because their light passes through a greater depth of atmosphere, and therefore encounters more turbulence, than light from stars at the zenith does. The degree to which stars twinkle varies from one night to the next, and seems to depend upon the prevailing weather. It is least pronounced when atmospheric pressure is high. However, no one has yet succeeded in establishing an infallible link between the weather and the degree of twinkling. Scintillation affects what astronomers call 'seeing'. If scintillation is very marked, seeing is said to be poor.

Planets tend to twinkle less than stars because of their greater apparent size. Nevertheless, they twinkle just like a star when they are seen near to the horizon.

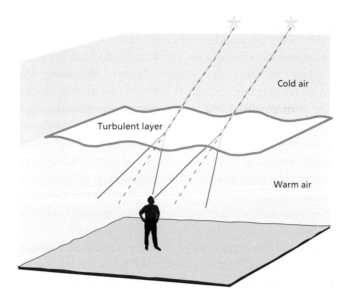

Figure 12.6 The reason stars twinkle is that starlight is refracted as it passes form cold air to warm air. Turbulence between the layers continuously changes the direction in which light refracts, causing the star to twinkle.

The discs of the Sun and the Moon are much too large for them to twinkle under normal circumstances. However, just before second contact, and just after third contact during a solar eclipse (see section 10.5), faint bands of shadow can be sometimes be seen racing across the ground in the area from which the eclipse is seen. These shadow bands are due to the combination of atmospheric turbulence and the narrow sliver of the almost-eclipsed Sun. The twinkling of stars and planets would also cause shadow bands if their light were brighter. Indeed, it has been claimed that, under the right circumstances, it's possible to see shadow bands on a white surface illuminated by a bright crescent Venus.

12.7 Seeing in the dark

Your eye is very far from being a perfect optical instrument. The limitations inherent in the construction and materials of even the most perfect human eyes place a limit on what you can expect to see when stargazing with the naked eye.

As has already been explained, only the very brightest stars shine with enough light to activate the colour sensitive cones of the eye, and all but the brightest stars can be seen only after your eyes have become accustomed to the dark. For dark-adaptation to occur, the light-sensitive rods must become active. This is a matter of chemistry. Rods contain a material called rhodopsin, which changes its chemical structure when it absorbs light. This change initiates the visual process. However, the rhodopsin must return to its original chemical state before it can absorb more light. Rods are about one million times more sensitive to light than are cones and hence, during the day, or in artificial illumination, they are over-stimulated. This renders them incapable of registering the light that falls on them. They require some 20 to 40 minutes of complete darkness to become fully sensitive to light again, though this varies from one person to another. Vitamin A, present in carrots and cod liver oil, improves the eye's sensitivity, whereas nicotine and alcohol prolong the time necessary to achieve full dark adaptation.

Given ideal 'seeing', once you have achieved dark adaptation you can expect to see as many as 2000 stars from any particular point on the Earth. Nevertheless, rods are so sensitive to light that even starlight is bright enough to reduce their sensitivity to some degree. This means that the eye can't achieve the theoretical limit of sensitivity when looking at a star-filled sky. However, if all but a tiny portion of the sky is blocked off by, say, looking at it through a small aperture from within a dark room, it is possible to improve the eye's sensitivity. This may be one reason why ancient astronomers used sighting tubes – long tubes without lenses. Indeed, use of these may well be the origin of the claim that it is possible to see stars during the day from the bottom of a deep mine shaft. The idea that this works is, of course, preposterous since it does nothing to reduce the brightness of the sky around a star, which renders it invisible in the first place. At night, however, the technique does enable very faint stars to be seen for the reason given above.

Although stars are not normally visible to the naked eye during daylight, under the right conditions it is sometimes possible to see Sirius, the brightest star, just after sunrise or just before sunset. An amateur astronomer, C. Henshaw, was able to see Sirius with the naked eye on several mornings up to 20 minutes after sunrise. At the time he made his observations, Sirius was 90° east of the Sun, and so almost overhead at sunrise. The observation was made from an elevation of 1100 m, 18° south of the equator in exceptionally clear skies. Y. Perelman, a Russian scientist, noticed that first magnitude stars are visible at mid-afternoon from the summit of Mt Ararat, which is 5 km high. Finally, when the Sun is eclipsed, a few of the brightest stars are sometimes visible during a totality.

12.8 Peripheral vision

You may have noticed that, if you want to see a very dim star, you have to look at it out of the corner of your eye. If you look at it directly, it appears to vanish. This frustrating phenomenon is due to the way in which rods and cones are distributed in the retina.

In daylight we look directly at any object that we wish to see in detail. In this way its image falls on a small spot on the retina, which is directly behind the lens of the eye. This spot is called the fovea. It is smaller than a pinhead and contains only cones packed very closely together. The rest of the retina is made up mainly of rods with a sprinkling of cones.

If the amount of light from a star is insufficient to stimulate the cones in the fovea then the star will not be seen when you look at it directly. If you glance away from it, however, its image falls to one side of the fovea, onto a region that is rich in rods. It may then become visible due to the rods' greater sensitivity to light. Peripheral vision is most sensitive on the nasal side of the eye. In other words you should turn your eye away from the star in the direction of your nose. By doing this, you focus light from the star on the most light-sensitive part of your retina.

You will discover that this technique, known as peripheral, or averted vision, is of use in noticing all manner of faint phenomena after dark when your eyes have achieved dark adaptation, not just for picking out faint stars. Peripheral vision can't be used in daylight, or during twilight, when vision is due only to cones.

12.9 Why are stars star-shaped?

Few people are able to see stars as points of light without the use of some sort of optical aid such as a telescope. Galileo charmingly described the appearance of stars to the naked eye as 'fringed with sparkling rays' and surrounded by 'adventitious hairs of light'. The same is true of distant lights. It was always thus: from ancient rock carvings to Renaissance frescoes, stars have invariably been drawn with spikes; and the symbol for a star is . . . a star.

A clue to the origin of these streaks is that they are particularly marked at night when your pupils increase in size to cope with reduced levels of illumination. Unfortunately, an enlarged pupil allows more light to pass through the outer edge of the eye's lens. Ligaments attach the edge of this lens to tiny muscles that focus the eye by altering the curvature of the lens.

Unfortunately these ligaments distort the lens, and are the cause of the streaks that fan out from the image of any point source.

To demonstrate that the streaks are due to an increase in the diameter of the pupil, try the following experiment. Look at a distant light at night from inside a brightly lit room. If you turn off the lights in the room, you will notice that the streaks of light radiating from the distant light increase in length and number because the aperture of the pupil is automatically increased to cope with the sudden darkness. The increase in the size of the pupil as it gets darker is the reason why Venus looks much smaller and point-like when seen soon after sunset when the sky around it is still bright, but becomes star-like when seen in the night sky. You will also find that the pattern of streaks is not the same for both eyes, and that the spikiness increases as you get older. Squinting improves the image of the star in the sense that it makes it more of a pinpoint. The improvement to an image brought about by squinting is more convincingly demonstrated when looking at a crescent Moon.

Stars really are pinpoints of light as you can find out by following Leonardo's instructions:

> If you look at the stars without their rays (as may be done by looking at them through a small hole made with the extreme point of a fine needle and placed so as almost to touch the eye), you will see these stars to be so minute that it would seem as though nothing could be smaller.
>
> Leonardo da Vinci

12.10 Constellations

Constellations are areas of the sky within which the brighter stars have been grouped together to form distinctive patterns to make it easier for us to find our way around the sky. In the western tradition, this practice began with the Babylonians. Modern star maps divide the celestial sphere into 88 areas of differing shapes and sizes, within each of which is to be found a number of bright stars that have been linked together to make a pattern that can, with some effort of imagination on your part, suggest the shape of a fanciful figure. The boundary of each constellation has straight edges that are parallel to either celestial longitude or celestial latitude. Stars visible within the boundaries of a particular constellation, but which do not form part of its shape, can be identified by the constellation within which they happen to lie.

A group of stars that does not make up a constellation, but which is nevertheless distinctive enough to have been named, is known as an asterism. Among of the best-known asterisms are the Plough, or Big Dipper, the Square of Pegasus, the Summer Triangle (Vega, Deneb and Altair), the Pleiades, or Seven Sisters, and the Southern Cross (visible only from the southern hemisphere).

The most distinctive constellation in the sky is probably Orion. Furthermore, since Orion lies on the celestial equator everyone on Earth can see it. Most people, however, probably associate constellations with the zodiac. This is a collection of constellations that happen to lie on the ecliptic, i.e. on the Sun's apparent path through the sky during the year. They are known collectively as the zodiac because all but one of them has been named after a creature. The astrologers of ancient Mesopotamia divided the ecliptic into 12 constellations. Since that time, however, astronomers have made various adjustments to the original scheme and in modern maps of the sky the ecliptic passes through 13 constellations. The thirteenth constellation, which is not recognised in astrology, is Ophiuchus and lies between Scorpius and Sagittarius. Many of the zodiacal constellations are quite difficult to see in their entirety with the naked eye since they all contain several faint stars.

Although the constellations that make up the zodiac are of unequal size, the largest being Virgo and the smallest being Libra, astrologers divide the ecliptic into 12 equal sections. Astrologically speaking, therefore, the Sun takes approximately one month to pass through each of these sections and its position on the celestial sphere can thus be given approximately by stating the sign through which it is passing. This means that the zodiacal constellations that you see in the night sky are those through which the Sun passed six months earlier. For example, although Cancer and Leo are visible high in the midnight sky during February, the Sun actually passes through them in August. Although astronomers do not make use of the zodiac, informal skywatchers may find it convenient, if they wish to give the approximate position of a planet or the Moon, to refer to the zodiacal constellation in which the celestial body can be seen rather than using the system of celestial coordinates employed by astronomers.

The starting point of the zodiac was fixed by Hipparchus in the second century B.C. He established the convention that the spring equinox occurs as the Sun enters the constellation of Aries. However, because of the precession of the Earth's axis, the ecliptic also precesses and in the 2000 years since Hipparchus, it has moved on by some 30°. Thus the spring equinox no longer falls in Aries. It now coincides with the start of Pisces, but it is still known as the first point of Aries even though it should now be the first point of Pisces. From an astrological point of view, this means that the Sun reaches the edge of the constellation of Pisces on 21 March of each year. Astronomically, the first point of Aries is where Sun crosses the celestial equator from the southern to the northern hemisphere in spring. This is the vernal or spring equinox.

The table below lists some of the more conspicuous objects visible in the night sky.

Some interesting celestial objects

Name	Date of Culmination	Remarks
Milky Way Galaxy	Highest in summer and winter evenings	This is actually our 'home' galaxy and it forms a broad diffuse band across the sky from one horizon to the other passing through the constellations of Cassiopeia and Cygnus in the northern hemisphere and through Crux in the southern hemisphere.
The Pleiades	18 November	A distinctive and closely grouped collection of several stars, some six of which are easily visible to the naked eye (they are also known as 'the Seven Sisters'). The Pleiades is a star cluster (i.e. they are all in more-or-less the same place and were formed at the same time).
Great Nebula in Orion	16 December	Just visible as a fuzzy star below the 'belt' of Orion. Binoculars reveal its true identity: not a star but a nebula.
Great cluster in Perseus	31 October	A closely packed group of stars that looks like a fuzzy star to the naked eye.
Great Nebula in Andromeda	1 October	Not really a nebula though to the naked eye it might well be one. Actually it is the only galaxy (apart from our own) that can be distinctly seen from the northern hemisphere with the naked eye. It looks like a faint patch of light. Use binoculars for a better view.
The Coal Sack in Crux	2 August	A dark gap in the band of the Milky Way that is due to the presence of dust between the Sun and it. Use binoculars for a better view.

Some interesting celestial objects (*cont.*)

Name	Date of Culmination	Remarks
Mizar and Alcor (known as the Horse and Rider)	The Dipper is circumpolar and so never sets at high northern latitudes	The pair of stars in the Big Dipper's handle (see figure 12.4). The horse is Mizar and the rider is Alcor. They are separated by 11 minutes of arc and so should be visible as separate entities to the naked eye.
Algol	6 November	Algol is a variable star: i.e. its brightness changes periodically. Such stars are common, but Algol is unusual in that its brightness varies visibly from 2nd magnitude to 4th magnitude for 4.5 hours once every 2.5 days.

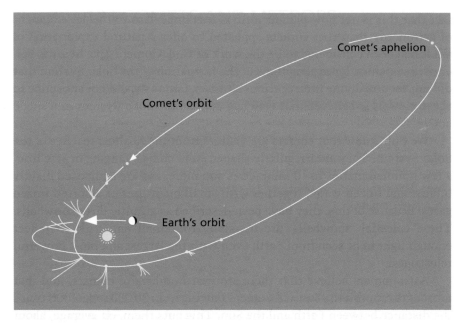

Figure 13.1 Orbit of a typical short-period comet. The comet is invisible until it begins to warm up as it approaches the Sun. As it grows warmer, it sheds gas and dust, which form the characteristic tail that we all associate with comets. This can grow to vast dimensions but is so insubstantial that it is blown away by the solar wind, so that it always points directly away from the Sun, as shown in the diagram.

through a mixture of reflection and fluorescence. Dust in the coma reflects light while the gases absorb ultraviolet radiation and fluoresce. Comets are the only objects in the Solar System, apart from the Sun itself, that are luminous, i.e. glow with their own light. Nevertheless, a comet owes its naked-eye visibility principally to the light reflected by the dust in the coma. In other words, the dust trail is considerably brighter than the gaseous one.

As an active comet nears the Sun, material from the coma is swept away because of the combined effect of the solar wind and the Sun's radiant energy. The resulting trail of dust and gas is known as a tail and can reach phenomenal proportions. Some comets have had tails that were more than 1 A.U. long. The proportion of dust to gas in the tail depends on the composition of the comet, which varies from one comet to another. The gaseous trail always points directly away from the Sun and is therefore fairly straight. However, since the gas is ionised by ultraviolet radiation, it is electrically charged and so is affected by the Sun's magnetic field. Variations in this magnetic field can cause kinks in the gaseous tail. The dust tail is noticeably curved because the ejected grains of dust lag behind the comet.

Disappointingly, most comets never grow bright enough to be seen with the naked eye or, indeed, with binoculars. Even those that do become visible to the naked eye often appear as little more than a faint smudge of light. Only the brightest comets, such as Hale-Bopp during March and April 1997, sport a distinct tail visible to the naked eye. Colour photographs of comets show that the gas tail is blue and the dust tail is yellowish or orange. To the naked eye, however, a bright comet typically appears as tiny bright core surrounded by a diffuse aureole with a distinct, though nebulous, tail. The bright core is the inner portion of the comet's coma and is not the nucleus, which is several hundred times smaller than the coma and therefore much too small to be seen at a distance even with a telescope. All parts of the comet appear white to the naked eye though, under favourable conditions, it is sometimes possible to notice the colour of the tail with binoculars.

Like all celestial objects, the brightness of a comet is given in terms of its apparent magnitude. However, a comet visible to the naked eye is an extended object and so its surface brightness may be quite low even when its stated magnitude is on a par with the brightest stars. To see that this is so, when there is next a bright comet, look through binoculars at a star of similar magnitude to the comet, and defocus its image until it is the same apparent size as the comet. The size of the image of the comet does not change much when it is defocused. Notice that the surface brightness of the defocused star has greatly diminished, whereas that of the comet is little changed. You can also reverse the process: by comparing the defocused image of a star with that of the comet you can make an reasonably good estimate of the comet's magnitude if you know the star's magnitude. See section 12.4 and 12.6 for a more detailed explanation of the relationship between surface brightness and magnitude.

The initial events that determine a comet's orbit occur so far away that they are obviously beyond the limits of any of our instruments, and so comets are among the most unpredictable of all astronomical phenomena. Each year at least 5–10 comets approach the Sun. Frequently, it is an amateur astronomer dedicated to the discovery of comets who makes the initial sighting. Comets are named after the person or persons who make the discovery.

Comets that put on a good show – i.e. become visible to the naked eye – share the following characteristics.

- The nucleus is large (at least 10 km across) and very active, i.e. material should sublimate from as much of the surface as possible.
- Either the comet's closest approach to the Sun (its perihelion) coincides

with a close approach to the Earth or the comet passes close to the Earth (so that it appears large but dim) or close to the Sun (smaller but brighter).

- The orbital plane of the comet lies well away from the ecliptic so that when the comet is close to the Sun, it will be visible above the horizon in the night sky.
- With the exception of Comet Halley, it is not possible to predict when a bright comet will appear.

All comets have highly eccentric orbits. This means that their perihelion distance is a fraction of their aphelion distance. (Aphelion is the most distant point from the Sun in the orbit of a body around the Sun.) Hence most comets take several thousand years to return to our skies because their orbits are such that, after making their closest approach to the Sun, they travel far away from it. A few comets, however, are manoeuvred into relatively small orbits through gravitational interactions with the planets, particularly Jupiter. A comet which completes its orbit of the Sun in under 200 years is classed a short-period comet (S.P.C.), which is why many S.P.C.s are seen once in a lifetime. Comet Halley, which has been returning to the inner Solar System at regular intervals since before the time of Christ's birth, is the most celebrated S.P.C., and owes its fame to its brightness rather than the fact that it was the first comet to be identified as an S.P.C. Being bright, it is visible to the naked eye, which is why its periodic return to the sky has been recorded century after century. When Edmond Halley, in whose honour the comet was named, examined comet records, he was able to work out that this particular comet had been seen many times at intervals of 75 years. He predicted that it would return in 1758, an event that he did not live to see for himself.

It has been estimated that a comet loses about 1% of its nucleus when in the Sun's vicinity. Generally speaking, therefore, short-period comets are intrinsically dim because they have grown smaller and less active because of their repeated passage close to the Sun. Comet Halley is the only S.P.C. that regularly reaches naked-eye visibility. It follows that, with the exception of Comet Halley, all bright comets are unexpected visitors to the inner Solar System, and are seen once only since their orbital periods are measured in several hundreds, if not thousands, of years.

People who have never seen a comet imagine that it moves rapidly across the sky rather like a meteor. But, since a comet orbits the Sun, and is usually very far from the Earth, its position in the sky changes gradually from one night to the next, and it may remain visible for several weeks, if not months. Occasionally, a comet passes so close to the Earth that it is possible to notice its motion across the sky from one hour to the next with the naked eye. This motion is even more obvious seen through binoculars. A recent example of a comet having a noticeable real motion was comet Hyakutake, which passed within 0.1 A.U. of the Earth in March 1996.

Short-period comets do not last forever. There are several ways in which they eventually vanish from the sky. In the first place, since a comet sheds material on each return to the Sun, eventually what remains may be no more than a trail of dust. These tiny fragments continue to orbit the Sun long after they leave the nucleus, and are the source of meteors. Comets are the major source of meteor showers, many of which are associated with active comets. Comet Halley is responsible for two distinct meteor showers: Eta Aquarids in May, and Orionids in October. Comets may also contribute to the dust responsible for the zodiacal light (see section 13.5). Another way in which a comet may cease to be active is for its surface gradually to become covered by a thick layer of dust, preventing the icy core from warming up and spouting gas. Gravity is yet another cause of cometary demise: its tidal effects may be large enough to cause a comet to break up into smaller pieces. Such was the fate of comet Shoemaker-Levy which broke up because of Jupiter's gravitational pull. The fragments of this comet crashed into Jupiter over a number of days, the only collision between a comet and a planet that has been witnessed. Finally, some comets collide with the Sun.

13.2 Meteors

The Earth in its orbit about the Sun does not move through a pristine void; it battles its way though large numbers of tiny asteroid fragments and vast amounts of dust, much of which has been shed by active comets. Collectively these particles are known as meteoroids. Most meteoroids resemble granules of freeze-dried coffee in both consistency and density, and range in size from a microscopic speck through to a pea. There is no defined upper limit to the size of a meteoroid, though meteoroids larger than a few centimetres are rare. An object several tens of metres across or more would be called an asteroid.

A meteor is the streak of light that is seen when a meteoroid plunges through the atmosphere. Meteoroids enter the atmosphere at velocities between 11 km/s and 72 km/s relative to the Earth. The reason why there is an upper limit to their velocity is because all meteoroids are in orbit around the Sun. Think of them falling towards the Sun from different heights.

Those that come from furthest away, from the edge of the Solar System, will reach the Earth's orbit with the greatest speeds. Meteoroids that orbit close to the Earth will enter the atmosphere with the lowest speeds.

It only takes a fraction of a second to convert the kinetic energy of a meteoroid into thermal energy as it plunges through the atmosphere at altitudes of between 80 km and 130 km above the Earth's surface. The brightness of the resulting meteor depends on the mass and velocity of the meteoroid. Even a meteoroid no bigger than a grain of sand can heat up to the point where it ablates, or burns up, creating a mass of extremely hot, glowing gas. Because of the persistence of vision, this is seen as a streak of light, which is known as a meteor.

You have only to spend a few hours out of doors on a clear night, even in a city, to catch a glimpse of at least one or two meteors as they flash briefly across the sky. On most nights, however, the tally of meteors is disappointingly small. If the Earth is colliding with huge numbers of meteoroids why isn't the sky lit up with a continuous celestial fireworks display? There are several reasons why we usually see only a few meteors. In the first place, the brightness of a meteor depends on its mass and velocity. Most meteoroids are so small that, despite their high velocity, they don't possess enough kinetic energy to heat up to the point where they ablate and glow. Secondly, although a meteor is an atmospheric phenomenon, its visibility is subject to many of the vicissitudes which bedevil stargazing: the degree to which your eyes have adapted to darkness, light pollution, atmospheric extinction and cloudy skies.

It has been estimated that many millions of meteoroids rain down on the Earth night and day. From the ground we see only a tiny fraction, about 0.5%, of the entire atmosphere that encircles the globe, and within which meteoroids burn up, and hence we see only a fraction of all meteors. Most of the meteors we see are almost overhead and relatively close to us, and are only slightly dimmed by atmospheric extinction. A meteoroid moving through the atmosphere just above the horizon is about 10 times further away than a meteor directly overhead, and is seen through the dense, hazy layers of the lower atmosphere. Thus, the mere fact that it is seen implies that it must be considerably brighter than the average meteor seen overhead. As for the rest, they are either too small, or are travelling too slowly to heat up to the point where they become visible. Very, very rarely a meteor is bright enough to be seen in daylight; these are a fertile source of U.F.O. sightings.

The total mass of meteoroids entering the atmosphere every day is estimated to be several thousand kilograms. Only the largest, densest meteoroids survive the intense heat generated as they plunge through the upper atmosphere, so few meteoroids reach the ground. Those that do are known as meteorites. Meteorites tend to be composed of either a nickel–iron alloy, or stony minerals, and are probably fragments of asteroids. Extraordinarily, a tiny handful of meteorites appear to have come from the Moon and Mars. These are fragments ejected from the surface of those bodies when they were themselves struck by large bodies, possibly asteroids. The largest known meteorite, the Hoba West meteoroid, named after the place where it lies in South Africa, is composed of iron, and has a mass of approximately 60 000 kg – too large to be moved. Most meteorites are far smaller than this, many no bigger than a small pebble.

Very bright meteors are known as fireballs. A fast-moving meteoroid the size of your fist can glow brightly enough to be seen in daylight. Occasionally a fireball will disintegrate and fall to Earth in fragments, each of which may itself glow brightly. Such fireballs are also known as bolides. Some fireballs leave persistent trails in their wake, known as trains, that are visible to the naked eye. These trains are due to electrification of air as the meteoroid passes through it. Meteor trains may last for several seconds or more.

The brightness of a meteor is assessed on the same scale as that used for stars and planets. It can be determined visually by comparing it to a star of similar brightness. This is something that can be done successfully only by someone who is already familiar with the night sky, though, of course, you can achieve such knowledge if you are sufficiently interested in doing so. Fireballs are meteors with a brightness greater than magnitude -3, i.e. almost as bright as Venus at it brightest.

The noise of a large meteoroid as it disintegrates is not usually audible from the ground because of the great altitude at which it usually breaks apart. The tenuous nature of the atmosphere at an altitude of 100 km means that sound-waves carry so little energy that they are inaudible. In any case, it would take several minutes for the sound to reach the ground. Nevertheless, some people have reported hearing buzzing, crackling and hissing sounds when they have seen very bright fireballs. It has been suggested that these noises may be due to very low-frequency radio waves emitted by the ablating meteor, which produce an audible sound when absorbed by objects at ground level. Some fireballs disintegrate a few kilometres above the ground and it is possible that the noise of these explosions is directly audible.

The occasional meteor that flashes across the sky on any clear night is known as a sporadic meteor. Generally speaking, sporadic meteors are two to three times more frequent in the early morning an hour or two before

(a)

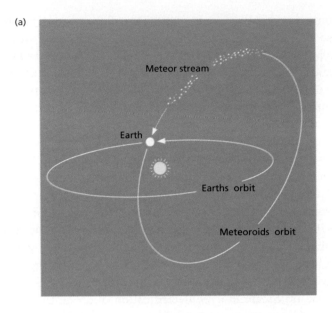

(b)

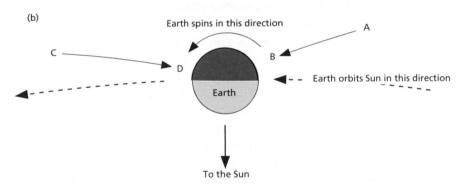

Figure 13.2 (a) Meteoroids are strung out along an orbit. If the orbit of the stream of meteoroids intersects that of the Earth, a meteor shower will be seen each time the Earth passes through the point of intersection.

(b) Generally speaking you are more likely to see a bright meteor after midnight. Meteoroids that orbit in the same direction as the Earth (from A to B) enter the atmosphere before midnight. At the same time, since they are catching up with the Earth, their relative velocity through the atmosphere is low, and they don't get as hot as meteoroids that enter the atmosphere after midnight. These meteoroids (from C to D) enter the Earth at large relative velocities because they are travelling in the opposite direction to the Earth in its orbit about the Sun.

sunrise than at any other time of night. This increase is due to the combined effects of the Earth's orbital and rotational motion. Both these are anti-clockwise viewed from the North celestial pole. The Earth is thus orbiting the Sun in the direction of the early morning sky and away from the direction faced by the evening sky. Meteoroids that orbit the Sun in the same direction as the Earth can enter the Earth's atmosphere only before midnight. The slowest of these will not be travelling fast enough to catch up with the Earth, while those that do, enter the atmosphere at low relative velocities and therefore may not produce a bright meteor. On the other hand, the morning side of the Earth sweeps up meteoroids. Furthermore, meteoroids orbiting the Sun in the opposite direction to the Earth enter the atmosphere at large relative velocities and are the cause of bright meteors after midnight.

At certain times throughout the year there are noticeable increases in the number of meteors. These events are called meteor showers, and are due to dust ejected by comets. As we have seen, a comet ejects dust to form a characteristic tail when it passes close to the Sun, i.e. when it is at perihelion. The tail is much wider than the parent comet, and over time, because of the gravitational tug of planets, gets wider and longer, forming a so-called meteor stream. Eventually, a meteor stream may form a complete loop of dust circling the Sun along the orbit of the parent comet. If the Earth passes through a meteor stream there is a possibility of a meteor shower. In fact, the Earth encounters many meteor streams in its yearly orbit, which is why meteor showers tend to occur at the same time each year. Meteor streams are typically several tens of times wider than the Earth's diameter and so the Earth may take several days to pass all the way through a stream.

The intensity of some meteor showers varies considerably from one year to another. A newly formed meteor stream will orbit the Sun close to the comet from which it was created because it will not have spread out along the whole of its orbit. Such a meteor stream will give rise to a meteor shower only if the comet reaches perihelion as the Earth passes through the orbit of the meteor stream. Spectacular displays of meteors that were seen in 1833 and 1966 were due to the Leonid shower that occurs between 14 and 20 November. Usually, this shower produces few bright meteors. But occasionally, in cycles of 33 years, it gives rise to quite unforgettable displays. However, the Leonid showers of 1998/99 were disappointing, and the number of meteors seen was average for this shower.

Although the particles that make up the stream move more or less parallel to one another, perspective makes it appear to someone on the ground as if meteors are radiating from one region of the sky. This region is known as the radiant of the particular shower. Indeed, a shower gets its name from

Figure 13.3 Meteor trail. A meteor blazes briefly across the sky, and its reflection in a foreground lake is caught by the camera. (*Photo* Pekka Parviainen)

the constellation within which its radiant lies. For example, the Leonid shower has its radiant in the constellation of Leo. The length of a meteor trail is affected by perspective: the meteors closest to the radiant will appear shorter than those further from the radiant.

Obviously, the higher up in the sky the radiant is when the Earth encounters the stream of meteoroids, the better the view that you will have of the resulting shower. However, you won't see the shower unless the radiant is visible in the night sky at your location.

Some major meteor showers

Name of shower	Best date	Remarks
Quadrantids	3 or 4 January	Peak is short and sharp and lasts for only 6 to 10 hours
Perseids	9–14 August	Maximum intensity 11 or 12 August Very intense shower
Orionids	18–24 October	Maximum intensity 20 October
Leonids	17 or 19 November	33-year cycle of peak activity
Geminids	10–12 December	Very intense shower

13.3 Artificial satellites

The first artificial satellite was Sputnik 1, launched by the Soviet Union on 4 October 1957. It orbited the Earth for about 90 days before plunging back to Earth and burning up in the atmosphere. Since then, thousands of artificial satellites have been launched.

There are several thousand artificial satellites in orbit around the Earth at any given time. Most of them are in orbits that are only a few hundred kilometres above the ground. However, only a handful are visible from the ground without binoculars or a telescope because most satellites do not reflect enough light to be seen with the naked eye. The best that you can expect to see without an optical aid is a bright point of light moving steadily across the sky without changing direction.

The visibility of a satellite depends on several factors, the most important of which are its size and height above the ground. As a rough guide, a

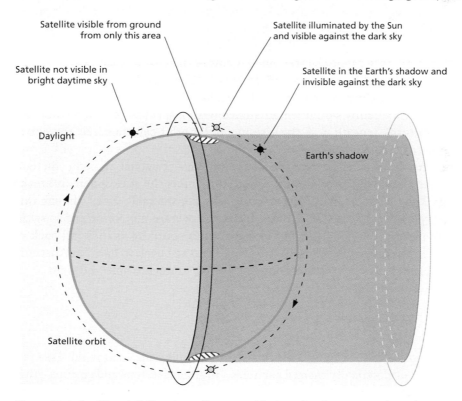

Figure 13.4 Satellite visibility. A satellite is visible for only a few minutes from the ground just after sunset or just before sunrise because it must be seen against a dark background while it is still illuminated by the Sun.

satellite 600 km above the ground would have to be at least 6 m across to reflect enough light to be seen by the naked eye. The brightest satellites such as the Space Shuttle can attain magnitude −1 and so are as bright as Sirius.

Something else that affects visibility is the time during which the satellite is favourably placed for observation. The orbital period of a satellite 400 km above the Earth's surface is approximately 90 minutes. However, being so close to the Earth, it only takes about 10 minutes to move from one horizon to the other, an angular distance of 180°. Hence this satellite will travel 10° in approximately 30 seconds. 10° is the angular width of your fist held at arm's length. In other words, a satellite 400 km above the Earth will move the width of your fist in half a minute. A satellite at twice this height – i.e. 800km above the Earth – takes approximately 15 minutes to travel from horizon to horizon and 50 seconds to travel one fist's width. If you are to see it during those few minutes when it is above the horizon, it has simultaneously to be illuminated by the Sun, and moving through a dark sky. These conditions occur only during twilight. The geometry of the situation is illustrated in figure 13.4. The long twilight at high latitudes during summer months provides the best conditions for satellite spotting. Note that a satellite cannot be seen once it enters the Earth's shadow.

Many satellites are in so-called polar orbits. In other words, they orbit the Earth in a north–south direction. Such orbits are favoured because the plane of a satellite's orbit remains fixed with respect to the stars while the Earth turns beneath it. A satellite in a polar orbit can thus 'see' the entire globe over a period of days.

Artificial satellites are the source of an occasional sporadic meteor. When the satellite's orbit decays, and it re-enters the atmosphere, it breaks up. The fragments, which are moving with considerable speed (though this is typically only one fifth of the speed of the average meteoroid and so satellite meteors are not as bright as cosmic ones), burn up as they fall back to Earth. Such was the fate of Skylab which crashed back to Earth in Australia on 11 July 1979, giving a large number of Australians a splendid sight.

13.4 Aurorae

The atmosphere is being constantly bombarded by the solar wind. This is a stream of electrically charged particles – mainly electrons and protons – that are emitted continuously by the Sun. As a result of the interaction between these particles and the air, the atmosphere literally lights up. And if it glows brightly enough we see an aurora.

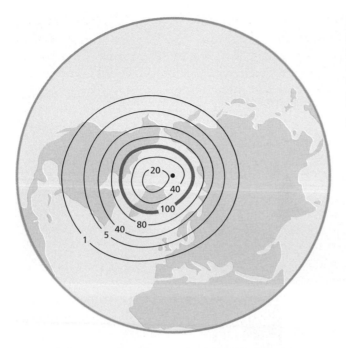

Figure 13.5 Map of the northern hemisphere showing the percentage of nights on which you can expect to see an aurora depending on where you are, weather permitting. The red dot marks the North Pole.

The first sign of an auroral display is likely to be a dim patch of light on the northern horizon. Most auroral displays develop no further. Occasionally, however, the initial glow brightens and grows. It begins to pulsate and within a few minutes the whole sky may be filled with multi-coloured bands and streaks and swirling curtains of light.

Since the solar wind is unrelenting, you might expect to see aurorae more or less continuously. There are, however, several factors that determine when and where aurorae will occur. In the first place, the intensity of the solar wind is not constant. It is particularly affected by an 11-year cycle of solar activity. At the peak of the 11-year cycle, auroral displays are both bright and widespread. The last maximum occurred in 1989 and lasted into 1991. The most intense displays are bright enough to wash out stars and even to cast shadows. At other times they may not be bright enough to be noticeable, at least not to anyone who is unfamiliar with what to look for.

The second major factor that determines these displays is the orientation of the Earth's magnetic field. This tends to confine most aurorae that occur in the northern hemisphere to a comparatively narrow band some 20° south of the Earth's north magnetic pole. And since this magnetic pole is at present to the west of the geographic pole, the band of maximum auroral

Figure 13.6 Aurorae. The appearance of an aurora depends on where it is relative to the observer. It looks like a curtain in the top photo because it is far away. In the photograph on the right the observer is looking up at the aurora because it is close by. (*Photos* Pekka Parviainen)

activity passes over northern Canada and misses northern Europe. The map shows the expected average frequency of auroral displays at different latitudes. In the British Isles, this can vary from 30 per year in Scotland to 5 a year in southern England.

The auroral displays seen in the northern hemisphere are called the Aurora Borealis; those seen in the Southern Hemisphere, the Aurora Australis. These should be regarded as one and the same phenomenon. Aurorae will occur in both hemispheres simultaneously during peaks of solar activity. Indeed, at such times, the displays may stretch from the poles to very nearly the equator. Nevertheless, generally speaking, few people in the southern hemisphere see aurorae because the band of maximum southern auroral activity passes over Antarctica and the uninhabited oceans that lie between Antarctica and Australia.

Aurorae come in a variety of shapes and colours. Characteristic shapes include small isolated patches, arcs, draped bands, rays and veils. Their colour ranges from white, when the aurora is not bright enough to stimulate the colour-sensitive cone cells in the eye, through to blue, green and red as well as combinations of these colours, such as magenta. The colours are characteristic of particular elements when they have been subjected to large electrical fields.

The colour of an aurora is also an indication of the height at which it occurs. Those that are formed at a height of between 100 km and 300 km tend to be green, whereas those that occur between 300 km and 1000 km tend to be red. Some reports speak of aurorae occurring very close to the ground so that they partially veil high mountains. Expert opinion tends to dismiss such sightings as visual illusions because of the difficulty in explaining how the aurora can form in the dense air of the troposphere.

A similar controversy surrounds the question of whether aurorae produce sounds. There have been innumerable reports of rustling, hissing, swishing or cracking noises accompanying auroral displays. But not everyone hears these sounds, so the matter remains unresolved. It may be that, like sounds that are sometimes reported to accompany bright meteors, auroral sounds are due to ultra low-frequency radio waves.

13.5 Zodiacal light

The vast quantity of dust ejected by comets that sweep by the Sun, and which is the source of many of the sporadic meteors that are occasionally seen flashing across the night sky, remains in orbit long after the parent

URBANDALE PUBLIC LIBRARY
3520 86TH STREET
URBANDALE, IA 50322-4056

Figure 13.7 The zodiacal
light appears as a faint wedge
of light rising from both
eastern and western horizon
along the ecliptic when the
Sun is below the horizon.

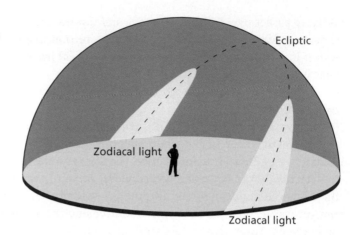

comets have ceased to exist. Collectively this dust scatters enough sunlight
to be visible as a faint glow in the night sky. The name given to this glow is
the zodiacal light, or false dawn.

Under favourable conditions, the zodiacal light may be seen from the
Earth's surface as a very faint, colourless wedge of light extending some 20°
or 30° along the horizon, and 30° or 40° into the night sky, along the ecliptic from the point on the horizon at which the Sun has recently set, or at
which it is about to rise. It is so faint that it is not visible even during astronomical twilight. The Sun must thus be at least 18° below the horizon if you
are to see it at all. Even then, it frequently goes unrecognised because at its
brightest it is no brighter than the Milky Way, for which it is often mistaken.
It is also sometimes taken for the glow of a distant town. However, it is possible to distinguish the zodiacal light from these because the Milky Way and
zodiac lie in different planes and the glow of town lights are not colourless.
It has been suggested that increasing levels of artificial illumination in
Europe and North America means that inhabitants of these regions are
unlikely ever to see the zodiacal light again – at least not from heavily pop-
ulated localities.

Another factor affecting its visibility is that the longest axis of the wedge
of light lies along the ecliptic. This orientation is to be expected because this
is the plane in which the interplanetary dust, which is its cause, orbits the
Sun. Because the dust lies in the plane of the ecliptic it follows that in northern latitudes, for example. North America and Europe, the most favourable
orientation of the zodiacal light to the horizon occurs twice a year: in
the evening around the time of the spring equinox and in the morning at the
autumn equinox. At other times of the year at these latitudes, the axis of the

zodiacal light lies too close to the horizon for the observer to be able to see it clearly. However, in the tropics, because of the permanently steep orientation of the ecliptic to the horizon, the zodiacal light is always favourably placed for observation either in the evening or in the morning. Some of the best views of it have been from the Peruvian Andes, which have the advantage of being near the equator, far from the light pollution of large cities and high above the horizon haze. Zodiacal light is also frequently seen at sea.

The characteristic shape and brightness of the zodiacal light is determined by the distribution of sunlight scattered by the interplanetary dust. This scatters most of the Sun's light in the forward direction, i.e. away from the Sun, and so the zodiacal light is brightest in the direction of the Sun, though visible only against a dark sky after sunset. Nevertheless, the belt of dust extends out towards Mars, and some sunlight is backscattered by dust between Earth and Mars, which, under exceptional conditions of darkness and atmospheric clarity, has been seen as a barely visible glow in the midnight sky. This is the elusive *gegenschein*, which is so faint that it is overwhelmed by the brightness of the Milky Way. Even experienced observers have failed to see it. The *gegenschein* is brightest at the antisolar point. This point, of course, sweeps around the celestial sphere because of the Earth's orbital motion. To stand a sporting chance of seeing the *gegenschein* you should choose a time when the antisolar point is (a) high above the horizon and (b) in a region of the sky well away from the Milky Way. The best time to look for it from the northern hemisphere is at midnight during October when the antisolar point lies in Pisces, a relatively dark constellation. It is claimed that under favourable conditions it is possible to see a band of light running across the sky from one horizon to the other, i.e. a combination of zodiacal light and *gegenschein*, which is known as the zodiacal band.

Appendix: Technical and practical advice for skygazing

Estimating distance

There are several ways in which we estimate how far something is from us. In most situations we rely on our acquired knowledge of the relative size of things and on the fact that objects that are close to us can block out parts of objects that are further away. However, these clues work only in circumstances and with objects with which we are familiar. Furthermore light, colour and shape can trick our eyes. An object may look nearer than it really is, if it is either uphill or downhill from us, if it is bright, if it is seen across water or snow or if the atmosphere is particularly clear (see section 1.5). On the other hand, in poor light or if the colour of the object blends with its surroundings, it may appear to be further away.

Even with knowledge of relative size, without binocular vision judging distance would be difficult in some situations. Binocular vision works by fusing two slightly different viewpoints of the same scene to give us a sense of depth. The spacing between our eyes, however, is small and so binocular vision is unreliable beyond a distance of some 30 m. You can check this for yourself by looking at an object that is more than this distance from you and then checking whether there is an apparent shift between object and background when you close one eye. This shift is known as retinal disparity.

For objects beyond 30m we make use of motion parallax. This is the relative motion that occurs between objects

Figure A1 Parallax. From A, the pole is seen before the house with the green door. From B it is seen before the house with the red door. The apparent shift in background due to a change in the observer's position is known as parallax.

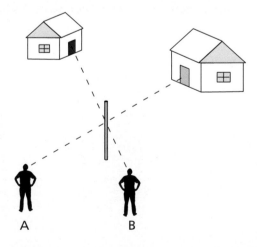

that are at different distances from us when we change the position from which we view them. Objects in the foreground change their positions with respect to their background much more than those that are further away from us do. This is readily noticed when looking out from a moving train or car. However, parallax between objects that are very distant from us is all but imperceptible (see also the remarks on the celestial sphere, section 8.3).

A primer on angles

When you are looking at optical phenomena such as a halo or a rainbow, you will sometimes find it useful to measure their angular dimensions. Fortunately, if all you want is an approximate value, you don't need to use an instrument because your hand can be used as a rough-and-ready cross-staff. The angle subtended at the eye by a fist held at arm's length is the same for almost everyone, namely 10 degrees. You can quickly confirm that this is the case by measuring the angle between the horizon and zenith using your outstretched fist. You will find that this is nine fists' worth, i.e. 90°. The diagram shows two further angles that can be easily measured in this way.

In some situations it is more convenient to measure angles in terms of radians rather than degrees. Radians are an alternative system of angular measure. One radian is defined as the angle subtended at the centre of a circle by an arc equal in length to the radius. An angle measured in radians is thus

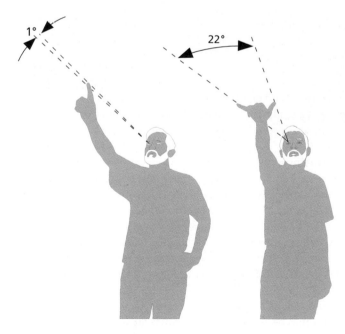

Figure A2 Measuring an angle with an outstretched hand. Your index finger held at arm's length subtends an angle of about 1°. If you stretch your fingers wide, the total angle will be approximately 22°, handy for locating where to look for ice halos. Your fist held at arm's length subtends 10°.

Figure A3 Relationship between arc length and radians. A radian is the angle subtended by a circular arc equal in length to the radius of the circle of which it forms a part.

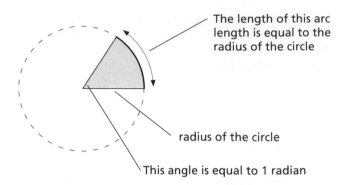

The length of this arc length is equal to the radius of the circle

radius of the circle

This angle is equal to 1 radian

the ratio of the length of the arc to the radius of the circle of which it forms a part. There are 2π radians in a circle because the circumference of a circle is $2\pi r$. 2π radians are thus equal to 360° and so 1 radian is 57°. Taking 1° to be $\frac{1}{60}$ radian (an acceptable approximation if you have to carry out the calculation in your head), the angular size of an object in degrees can be converted into radians to a good approximation by dividing it by 60.

Measuring angles in radians is useful because, if either the size of the object or its distance from the observer is known, small angles can be easily converted into distances or dimensions. For example, if an object 5 metres across subtends 1° (or approximately $\frac{1}{60}$ radian), it is 300 metres from the observer ($\frac{5}{300} = \frac{1}{60}$). This technique of estimating distance works best for small angles because the definition of the radian is based on an arc. When measuring an angle you can't allow for this curvature. For an object that subtends 30° (approximately half a radian), the discrepancy between the estimated arc length (i.e. not allowing for its curvature) and the true one amounts to 2%,

at 60° (approximately one radian) it is 5% and at 90° (approximately one and a half radians) it is 10%.

Binoculars

Binoculars are essential kit for anyone with an interest in nature. If you don't already own a pair, you may find the following notes useful in selecting a suitable pair.

Binoculars are really a pair of folded telescopes. Between the objective and the eyepiece there are prisms which reflect the light up and down the tube hence reducing its physical length while maintaining its optical length. Being compact, the effect of hand tremors is reduced. At the same time, since we normally use both eyes, binoculars preserve some illusion of depth (though, as you will have noticed, perspective is compressed. This is because the magnification brought about by the binoculars does not allow the images to subtend the same angles as they do with the naked eye).

Binoculars are available in many combinations of size and magnification.

Each pair is specified in terms of its magnification and diameter of objective (the lens nearest the object). For example, 7×50 binoculars give a magnification of 7 times and have objectives of diameter 50 mm.

When choosing a pair of binoculars consider the following points.

- *Diameter of objective*: the larger this is, the greater the amount of light gathered and so the larger the magnification that the image can undergo without loss of detail.
- *Magnification*: although large magnification is desirable because it allows one to see more detail, it becomes increasingly difficult to hold binoculars with a magnification greater than about 10 times steady without a means of support.
- *Exit pupil*: the exit pupil is the narrow pencil of rays that emerges from the eyepiece. Hold a pair of binoculars at arm's length and point them at a distant scene: the exit pupil is the image of the objective lens seen though the eyepiece. All the light that enters the objective is concentrated into the exit pupil. If binoculars are to work efficiently, all the light gathered by the objective should enter the eye. This can happen only if the exit pupil of the instrument does not exceed the pupil of the eye. Indeed, it is better if the exit pupil is smaller than the pupil of the eye to confine the light that emerges from the eyepiece to the central portion of the eye's lens. Note, also, that the maximum diameter to which the pupil of a dark-adapted eye can dilate diminishes with age. In the dark, a child's pupil

may dilate to 7 mm; a middle-aged eye may be capable of dilating to only 5 mm. The exit pupil of binoculars may be calculated by dividing the diameter of the objective by the magnification. For 7×50 binoculars the diameter of the exit pupil is 7 mm.

- *Field of view*: generally speaking, the larger the magnification, the smaller the field of view. Standard 7× binoculars have a field of view of approximately 7°, while for 10× binoculars the field of view is 5°. The field of view is usually given on the binoculars. Ultra-wide-field binoculars are also available. These can give a field of view of 10° to 12° with 7× magnification. A large field of view has advantages: breathtaking views of starfields at night and wider views in daylight.
- *Optical quality:* lenses should be multicoated, i.e. both surfaces of every lens should be treated to improve light transmission. Avoid zoom binoculars since these are usually a very poor compromise between magnification and aperture.
- *Portability:* if you are going to carry binoculars around with you, it is worth keeping in mind that those with large objective lenses are going to be bulky and, possibly, heavy.

The standard recommendation for astronomical observation is 7×50, principally on the grounds of its light-gathering capacity. However, in the light of the points made above you may decide that you would be better off with 7×35. These are much more portable, give the same magnification, while having an exit pupil better matched to the adult eye. In daylight, of course, there is no need for the greater aperture of the 7×50s.

A cloud primer

Meteorology, the science of weather, is not directly relevant to the topics covered in this book. Nevertheless, if you look at the sky you will inevitably see clouds, some of which are directly involved with optical phenomena.

The only way to learn how to identify clouds is to make a point of doing so. Whenever you are out walking, or simply pottering about, make a point of looking up at the sky, and try to identify the clouds present. This requires more than a cursory glance. You will often have to make decisions about the nature of what you see. Cloud identification is not a matter of blindly applying labels: you have to develop an eye for cloud types. There will be many times when you will be unsure what type of cloud it is that you saw, and you may only come to realise what it was sometime later when you see it again. There are many permutations for each cloud type so use a cloud atlas. If you are diligent, within a few weeks, at most, you will find that your knowledge of clouds will have improved enormously.

The modern system of classification is based on a cloud's shape, the idea for which was first put forward by an Englishman, Luke Howard, in 1803. In this system there are three basic cloud shapes: heaps (known as cumulus or cumuliform clouds), layers (stratus or stratiform) and fibres (cirrus or cirri-

form). What makes the scheme both useful and workable is that the appearance of a cloud is governed by the process that creates it. Particular processes result in particular shapes. Appearance is also an indication of consequences: certain types of cloud give rise to rain, others to thunder storms while yet a third type is associated with good weather. In addition, two of these cloud types can occur at different altitudes. The prefix *cirro* is applied to clouds that occur at the greatest altitude, while *alto* is for those that occur at intermediate altitudes. Rain-bearing clouds are given the prefix '*nimbus*' or '*nimbo*'.

The international system of cloud classification recognises 10 basic types of cloud. They are all listed in the table. Some clouds fit more neatly into this scheme than others. Cirrus and cumulus are easily identified. Others may present difficulties since one type of cloud can sometimes develop from another, and so intermediate forms occur, for example altostratus develops from cirrostratus. Frequently, more than one type of cloud is present in the sky.

Even the most casual observer can't fail to notice that clouds are constantly changing shape and size. Few individual clouds last for more than an hour. Hence you must resist the temptation to think of clouds as entities. A cloud is not a thing in the sense that, say, a mountain is. Rather, it is the manifestation of a complicated physical process that involves the condensation of a mass of moist air and its subsequent evaporation. If you keep an eye on a group of clouds over a period of a several minutes or more you can often form a good idea of what the otherwise invisible air is doing: for example whether it is ascending or descending, whether it is dry or moist, the direction of the prevailing wind, whether the sky is likely to remain overcast and the possibility of rain. Clouds, or the lack of them, are the only guide that someone on the ground, and without any meteorological equipment, has to what is happening in the atmosphere above us. And this knowledge, in its turn, provides clues to both short-term and long-term weather patterns.

Cloud types

Altitude	Heaped	Layered	Fibrous	Vertical
High level (cloud base 7 to 13 km)	Cirrocumulus (Cc)	Cirrostratus (Cs)	Cirrus (Ci)	Cumulonimbus (Cb)
Medium level (cloud base 2 to 7 km)	Altocumulus (Ac)	Altostratus (As)		Cumulonimbus
Low level (cloud base 0 to 2 km)	Cumulus (C)	Stratocumulus (Sc)	Cumulonimbus	
	Nimbostratus (Ns)	Stratus (St)		

Glossary

ablation   The erosion of a surface, say through melting, vaporising or weathering.

absorption   When an atom absorbs an electromagnetic wave, the energy of the atom changes and the wave ceases to exist. Depending on the nature of the atoms of which an object is composed, the energy of the absorbed wave may increase its temperature, bring about a chemical reaction or be emitted as electromagnetic energy of a different wavelength. Visible wavelengths that are not absorbed are responsible for the colour of the absorbing material: for example leaves and grass look green because they contain chlorophyll, a substance that absorbs the red end of the spectrum.

aerosol   Liquid drops or solid particles ranging in size from 5 to 50 microns (1 micron = 1 millionth of a metre). They are thus small enough to be suspended for long periods in the atmosphere.

airlight   Light scattered by molecules in the atmosphere. It is noticeably blue because of Rayleigh scattering.

albedo   A measure of the reflectivity of a surface expressed as the fraction of incident light that is reflected by the surface. The albedo of snow is almost 1 and that of lamp black is almost 0.

angular diameter   The angle an object subtends at the eye. The angle is enclosed by the apex of a triangle, the base of which is the object and the apex the eye of the observer.

antisolar point   The point on the opposite side of the sky from the Sun. The line joining this point to the Sun passes through the eye of the observer; the shadow of an observer's head always falls on the antisolar point.

apogee   The most distant point from the Earth in an orbit around it.

atmospheric optics   The general name for any optical phenomena within the atmosphere due to the interaction between sunlight, air and water in the form of drops or ice crystals.

A.U.   A.U. stands for astronomical unit, and is the average radius of the Earth's orbit about the Sun. 1 A.U. is 150 million kilometres.

aureole   A bright, colourless glow often seen around the Sun. It is due to forward scattering by aerosols in the atmosphere.

brightness   The subjective sensation of luminosity.

celestial equator   The projection of the Earth's equator onto the celestial sphere. The celestial equator divides the celestial sphere in two: the northern celestial hemisphere and the southern celestial hemisphere.

celestial sphere   The imaginary sphere of the sky centred on the Earth.

circumpolar   Because of the Earth's daily rotation, all stars appear to move across the sky in circles centred on the celestial poles. Stars that never disappear below the horizon are said to be circumpolar. For observers at either pole, all stars are circumpolar. With diminishing latitude, the number of such stars become fewer until at the equator no stars are circumpolar.

colour   The physical basis of colour is the frequency of light. Colour, however, is a sensation, and the relationship between the stimulus and what is perceived is a complex one. There are three types of receptor, collectively known as cones, in the retina, each of which is sensitive to a broad band of frequencies. There is a large overlap between each of these bands, but because the three types of cone have peak sensitivities in different parts of the spectrum, they are identified with long, medium and short wavelengths respectively. The signals from each type of receptor are combined together in such a way that the signal sent to the brain is a composite of the responses of the three types of receptor.

In fact, the word 'colour' is not an adequate description for how something appears to the eye. There is no unique definition of any particular colour and, as we all know, every colour comes in a bewildering variety of shades. To adequately describe the attributes of a light or a pigment we really need to refer to three qualities: its hue, its saturation and its brightness. Hue is the dominant colour, and is what most people understand by the word colour. Saturation refers to the purity of a hue. Mixing it with white light reduces the saturation of a particular hue. Brightness refers to intensity. Grey is just a less bright version of white. Black is the complete absence of light.

conjunction   An astronomical configuration in which the position of a planet appears to coincide with that of the Sun. i.e., seen from the Earth, the planet and the Sun lie in the same part of the sky.

cosmology   The science of the Universe as a whole.

culmination   The date of culmination is the day on which a particular star reaches its highest point in the midnight sky.

dazzle   Light scattered within the eye, making it difficult to see clearly.

diffraction   When waves encounter an object, a new set of waves is created which spread out from the object. This phenomenon is known as diffraction. Diffraction is most pronounced where the obstacle has the same dimensions as the wavelength of the waves.

dispersion   The colour separation that occurs when a beam of white light is refracted, say as it passes through a glass prism (see figure G1). The degree of refraction is inversely related to wavelength, so violet refracts more than red, with other colours in between. This spread of colours is the spectrum of white light.

eclipse   An astronomical event in which the shadow of the Moon falls upon the Earth or in which the Moon passes through the Earth's shadow. The word 'eclipse' comes from the Greek for 'failure' or 'to leave out'.

ecliptic   The plane in which the Earth orbits the Sun.

electromagnetic radiation   Energy that is transferred by means of electromagnetic waves. The term has become associated with the emissions of radioactive substances, though strictly speaking only one of these emissions can be classified as such, namely gamma rays. The full spectrum of electromagnetic radiation,

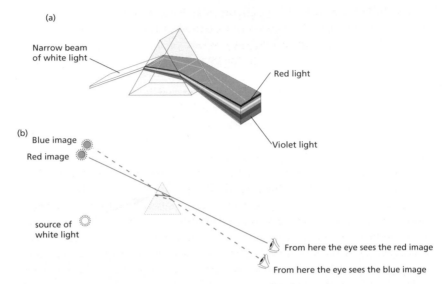

Figure G1 (a) When a narrow beam of white light passes through a prism it separates into a series of colours. This is the spectrum of white light. The colours in the spectrum are in the following order: red, orange, yellow, green, blue and violet. Indigo is sometimes said to be a seventh colour lying between blue and violet, though, in my experience, it is not possible to discern it.

(b) The lower diagram shows refraction through a prism. Looking through the prism at a source of white light the eye can't see the red and blue image at the same time. Curiously, although the blue end of the spectrum emerges from the prism below the red end, it is seen above the red when you look into the prism.

in order of decreasing wavelength, consists of radio waves, microwaves, infrared, visible light, ultraviolet, X-rays and gamma rays. Visible light itself has a spectrum, which runs from red to violet. (See figure G2).

elongation   The apparent angular separation between an astronomical object and the Sun as seen from the Earth.

extinction   The attenuation of a beam of light as it passes through a nominally transparent medium such as the atmosphere. Extinction is due to absorption or scattering or both.

frequency   The number of cycles per second performed by a vibrating object. One cycle per second is known as a Hertz. *See also* period, wavelength

galaxy   A galaxy consists of a vast number of stars (anything from one million to one million million within a single galaxy) together with huge quantities of dust and gas gathered together into a single system held together by mutual gravitational attraction. The Milky Way is our 'home' galaxy.

geocentric   Centred on the Earth. In a geocentric universe the Earth is at the centre of the Universe and the Sun, Moon, planets and stars all orbit the Earth.

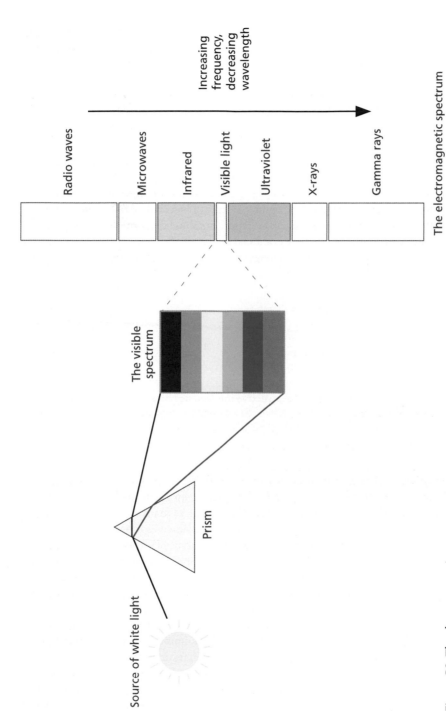

Figure G2 The electromagnetic spectrum.

gibbous   The phase of the Moon or a planet when it is between half and full phase.

haze   A brightening of the atmosphere due to sunlight scattered by a high concentration of aerosols.

heliocentric   Centred on the Sun.

heliacal   Rising or setting of a star or planet with the Sun.

illuminance   The amount of light per unit area per second illuminating a particular surface.

infrared   A form of electromagnetic radiation invisible to the eye, but perceptible as warmth by the skin. The Sun emits about half its radiant energy as infrared, and the other half as light.

interference   A phenomenon common to all wave motion. It occurs whenever two or more waves combine. The combination can be constructive or destructive. When waves arrive at a point in step they combine constructively, and when waves arrive out of step they combine destructively. If destructive, the wave appears to vanish, though its energy has actually been transferred elsewhere.

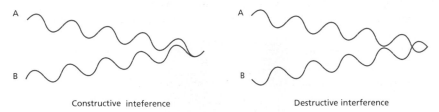

Constructive inteference   Destructive interference

Figure G3 Constructive and destructive interference of waves. If waves A and B are in step when they come to together, they combine to create a wave of larger amplitude. If they are out of step they cancel one another out.

ionisation   All electrical phenomena are due to the equal and opposite charges carried by the protons and electrons that make up all atoms. In the normal state an atom is electrically neutral since it has as many protons (which have a positive charge) as it has electrons (which are negatively charged). If an atom loses or gains electrons it is said to have been ionised. Losing electrons gives the atom an overall positive charge while gaining them makes it negative. Under normal circumstances it is not possible for the atom to lose or gain protons because they are bound up within the nucleus and therefore cannot be removed without totally destroying the atom.

irradiation   Light is scattered within the retina and so light from a bright source may stimulate retinal cells that lie outside the immediate area within which the source's image is focused. The result (irradiation) is that the image of the object appears larger than it really is. Irradiation also occurs in photographs.

light year   Interstellar distances are so vast that it is more convenient to measure them in terms of the distance covered by light in one year than in ordinary linear measure. One light year is the distance a beam of light can travel in one year. The distance between the Earth and the Sun is approximately 8 light minutes. To convert light years into kilometres multiply the velocity of light in a vacuum (300 000 km/s) by the number of seconds in a year ($365 \times 24 \times 3600$ seconds). This works out at 9 460 800 000 000 km. On the scale used to model the Solar System (Earth = 1 cm) this works out to be 7400 km. Using this scale, the nearest star (Proxima Centauri, 4.2 light years) is 31 000 km from the Solar System (half as far away again as Australia is from Europe).

luminance   A measure of the amount of visible light emitted per unit of apparent area of a source. Luminance is independent of the distance between source and observer because, although as you move away from the source the eye receives less light, the image of the source upon the eye becomes correspondingly smaller.

luminosity   The total amount of energy emitted per second by a radiant source such as a star.

lunation   A complete cycle of lunar phases from one new Moon to the next. The time taken for this is 29.5 days, the Moon's synodic period.

magnitude   A measurement of the brightness of a celestial object such as a planet or a star.

meteor   The brief luminous trail seen in the night sky when a meteoroid ablates as it passes through the atmosphere.

meteorite   A meteoroid that has survived ablation and reached the Earth's surface.

meteoroid   A fragment of rock, metal or dust in orbit about the Sun.

molecule   The smallest unit of a substance. Molecules are composed of atoms.

nebula   A cloud of interstellar dust or gas that may be either the material from which a star will be formed or the remains of an exploded star.

opposition   An astronomical configuration in which Earth lies directly between a superior planet and the Sun. The planet appears to lie on the other side of the celestial sphere from the Sun, i.e., seen from the Earth, the planet and the Sun lie on opposite sides of the sky, or 180° apart from one another.

orbit   The path of a planet about a star, or a satellite about a planet, due to the mutual gravitational attraction between the two bodies. The orbits of planets and satellites are elliptical, though in most cases the ellipses are very close to being circular.

parallax   The apparent shift in position in an object against a distant background when we alter our viewing point.

parsec   A unit of astronomical distance based on angular measure. 1 parsec is the distance at which the radius of the Earth's orbit subtends one second of arc. 1 parsec is equivalent to approximately 3.2 light years, or 30 000 million million kilometres, or 200 000 times the distance between the Earth and Sun. The nearest star, Proxima Centauri is 1.3 parsecs from Earth.

perigee   The closest point to the Earth in an orbit around it.

period   Time taken for an orbiting body to complete one revolution. *See* sidereal period and synodic period. Also time taken for an oscillation or wave to complete one cycle. *See* wavelength.

phase   In astronomy this is the amount of the illuminated surface of a celestial body that is visible from Earth.

photopic vision   Vision by cone cells when the level of illumination is high, for example in daylight. Cone cells are responsible for colour vision.

pigment   A material whose colour is due to the absorption of light. The colour of a pigment is that part of the spectrum that has not been absorbed by the chemical from which the pigment is made.

planet   A body that orbits a star and which is not hot enough to emit visible light. Planets are made visible through reflected light.

planitesimal   A geologically primitive type of astronomical body with a diameter up to several hundred kilometres which, it is believed, condensed from the solar nebula and from which the planets were formed.

precession   The axis about which a planet such as the Earth rotates does not point at a fixed point in space: it gyrates, slowly tracing out a closed path. It takes 25 800 years for the Earth's axis to precess.

quadrature   An astronomical configuration in which a planet or the Moon is 90° from the Sun as seen from the Earth.

reflection   All types of wave are reflected when they meet a surface. However, not all the incident wave energy is reflected: some always passes into the medium of the reflecting surface where some or all of it may be absorbed. In all reflections the reflected wave makes the same angle with the normal (a line perpendicular to the reflecting surface) as the incident wave does. This relationship is known as the law of reflection. A smooth surface gives rise to a specular (mirror-like) reflection, a rough one brings about a diffuse reflection.

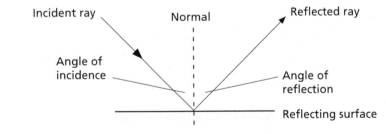

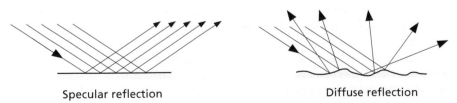

Specular reflection   Diffuse reflection

Figure G4 Reflection. The angle of incidence is always equal to the angle of reflection, even when the reflecting surface is rough.

refraction   When a wave passes from one medium to another its velocity changes. However, since the wave must keep step with itself, its frequency remains constant and hence, to compensate for the change in velocity, its wavelength changes. The net result is that it changes direction on entering the new medium.

The relationship between the angle of incidence and refraction (i.e. the law of refraction) is not as straightforward as that for reflection. A wave follows the path of least time from the source, through the refracting medium to any point on the other side of this medium. The refractive index for any pair of media is equal to the ratio of the velocities of the wave in each media.

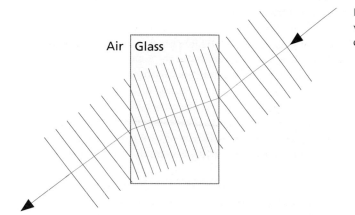

Figure G5 Refraction of a wave occurs when its speed changes.

Air | Glass

scattering   When an electromagnetic wave is scattered by an atom some of its energy is radiated away from the original direction in which the wave was travelling. The amount of energy that is scattered, and the volume over which it is scattered, depends on both the size of the particle and the wavelength of the wave.

Although scattering can and does occur in solids, liquid and gases, its effects are most pronounced for matter in a dispersed state, i.e. where the constituent particles are small and widely separated from one another. Hence scattering is important in determining the appearance of some gases and of hazes of very small particles such as huge molecules, tiny dust particles and tiny drops of water (e.g. those making up mists and fogs). If the particles are small molecules, scattering can bring about a change in the colour of a beam of white light by scattering some wavelengths more strongly than others.

The amount of energy scattered when an electromagnetic wave encounters a very small particle (several hundreds of times smaller than the wavelength of light) is proportional to the square of the volume of the scattering particle and inversely proportional to the fourth power of the wavelength of the radiation.

The wavelength of red light is about 1.7 times that of blue light and so a given particle will scatter approximately 10 times as much blue light as red light. Hence when white light passes through a medium made up of a large number of very small particles, scattering redistributes the wavelengths that make up a beam of white light. The transmitted beam becomes increasingly redder because more blue light is scattered out of it than red light. This scattered blue light is seen most clearly when the beam is seen perpendicularly to the direction in which it travels. Such scattering is now known as Rayleigh scattering or selective scattering.

Although Rayleigh scattering is most pronounced on the scale of atoms and molecules, even particles that have a diameter of 0.3 μm (approximately one tenth the wavelength of light) scatter light selectively. Larger particles also scatter light but do not do so selectively: in this case all wavelengths are scattered to an equal degree and so the scattering process does not, in itself, lead to colour effects. However, colours may be produced through secondary effects such as constructive and destructive interference. Scattering for large particles is mainly confined in the forward direction (i.e. in the direction in which the radiation is travelling), though there is also a degree of backscattering (i.e. light scattered in the direction of the source). *See also* absorption.

scotopic vision   Vision by rod cells under conditions of very low illumination. Rod are sensitive only to intensity and not to wavelength, and so such vision is monochromatic.

self-shadowing   This occurs when the brightness of an object that lies between an observer and the Sun is reduced because some or all of the object's surface which can be seen by the observer lies within the object's own shadow.

shadow hiding   An object at the antisolar point hides its shadow from the observer who consequently sees only its illuminated surface. Shadow hiding is the opposite of self-shadowing, and it is the cause of the increased brightness seen at and around the antisolar point on some surfaces.

sidereal period   The time taken for a celestial body to complete an orbit with respect to the fixed stars. The Earth's sidereal period is 365.25 days.

star   A large body composed mainly of hydrogen and helium that emits visible light and other forms of electromagnetic energy due to a nuclear reaction at its core. Stars differ from one another in mass and luminosity.

sublimation   When a solid becomes a gas without going through the liquid phase. For example, a lump of dry ice (solid carbon dioxide) will gradually shrink as it becomes a gas. It does not first form a puddle.

synodic period   The time taken by a celestial body to return to the same position in the sky. The synodic period can be calculated from the sidereal period with the following formulae.

For inferior planets $1/S = 1/T - 1/E$
For superior planets $1/S = 1/E - 1/T$

where S is the synodic period of planet

E is the Earth's sidereal period (= 365.25 days)

T is the sidereal period of planet

syzygy   The astronomical conjunction or opposition of orbiting bodies. For example, at a syzygy, the Earth, Moon and Sun are lined up with one another and an eclipse may occur if the Moon is at or near one of its orbital nodes. The term 'syzygy' is derived from the Greek for 'union' or 'coupling'.

tectonic   The process responsible for the deformation of the Earth's crust or changes due to this.

temperature inversion   The normal situation in the atmosphere nearest the ground (the troposphere) is that the temperature of the air decreases with height all the way to the tropopause, at the rate of approximately 1 °C per 100 m. This is the normal lapse rate. However, there are circumstances under which, at a certain altitude, the temperature of the air actually begins to increase. This situation is known as a temperature inversion. The tropopause is a permanent temperature inversion that separates the troposphere from the stratosphere. This inversion prevents convection clouds from rising into the stratosphere.

terminator   The boundary between day and night.

turbulence   The flow of a fluid over a surface is smooth at low velocities. However, as its velocity increases, the flow breaks up into eddies and becomes turbulent.

visual acuity   Some people have better eyesight than others. It is generally agreed that human eyesight is capable of resolving details that have an angular size of one minute of arc. This means that, under ideal conditions, someone with perfect eyesight should be able to distinguish a circle of diameter 1 cm from a square of side 1 cm at a distance of 34 metres. Few people possess eyesight as good as this, and in any case, acuity drops off with age. Fortunately, suitably chosen eyeglasses go a long way to correcting the eye's deficiencies. Visual acuity also depends on the contrast between what one is looking at and its background. When this is not great, as in the case of the lunar seas, detail is lost to the eye.

visual range   The greatest distance at which an object can be unambiguously identified.

wavelength   The distance a wave travels in one cycle. In the case of a wave travelling across the surface of water, this is equal to the distance from one crest to the next.

zenith   The point in the sky directly above your head.

Further reading

These are books that I have found particularly useful, and to which frequent reference is made in the sources and notes. Those asterisked are still in print at the time of writing.

* Bohren, C.F., 1987, *Clouds in a Glass of Beer*, John Wiley. Subtitled *Simple experiments in atmospheric physics*. Pungent treatment of many perennial errors and misconceptions in the physics of atmospheric phenomena. Topics are approached through simple experiments which readers are encouraged to carry out for themselves. Abbreviated in the Sources and notes as *Clouds.*
* Bohren, C.F., 1991, *What Light Through Yonder Window Breaks*, John Wiley. Subtitled *More experiments in atmospheric physics*. Bohren continues the good work he began in *Clouds in a Glass of Beer*. Abbreviated in the Sources and notes as *What Light.*
* Greenler, R., 1980, *Rainbows, Halos and Glories*, Cambridge University Press. Particularly strong on halos, though, arguably, now surpassed by Walter Tape's *Atmospheric Halos* (see Sources and notes, section 7.1). Fine selection of illustrative photographs. Abbreviated in the Sources and notes as *Rainbows.*
Humphreys, W.J., 1964, *Physics of the Air*, Dover. An exhaustive account of atmospheric phenomena. The treatment is technical though not dauntingly so. Unfortunately out of print.
* Kaler, J.B., 1996, *The Ever-Changing Sky*, Cambridge University Press. A comprehensive text dealing with the fundamentals of positional astronomy.
* Können, G.P., 1985, *Polarized Light in Nature*, Cambridge University Press. Exhaustive, non-technical treatment of polarisation phenomena with particular emphasis on observation. Excellent selection of illustrative photographs. Abbreviated in the Sources and notes as *Polarized Light.*
Lynch, D.K. (ed.), 1980, *Atmospheric Phenomena*, W.H. Freeman. Articles from past issues of the *Scientific American* dealing with ice crystals, meteorological optics, thunderstorms and some nocturnal phenomena such as the aurora.
* Lynch, D.K. and Livingston, W., 1995, *Colour and Light in Nature*, Cambridge University Press. Excellent guide to a wide range of optical phenomena in nature. Abbreviated in the Sources and notes as *Colour and Light.*
Meinel, A. and Meinel, M., 1983, *Sunsets, Twilights, and Evening Skies*, Cambridge University Press. Particularly strong on abnormal twilight due to volcanoes. Excellent selection of illustrative photographs. Abbreviated in the Sources and notes as *Sunsets.*
* Minnaert, M., 1954, *Light and Colour in the Open Air*, Dover. This is the classic

book on meteorological optics. Covers a vast range of optical phenomena in an unassuming, though thorough, manner. It is a translation from the original Dutch of the first volume of three that deal with the physics of natural phenomena. The series was published in Holland under the title *De natuurkunde van't vrije veld* by N.V.W.J. Thieme & Cie in 1968. A much more recent translation of a 1974 Dutch edition is entitled *Light and Colour in the Outdoors* (Springer Verlag, 1993). This contains some material not in the Dover edition and is illustrated with colour photographs. There are careless misprints in some of the calculations. All references in the notes are to the Dover edition. Abbreviated in the Sources and notes as *Light and Colour*.

* Ottewell, G., 1989, *The Astronomical Companion*, Astronomical Workshop, Furman University, Greenville, S.C., 29613, U.S.A. Probably the best single volume guide to the sky available for anyone interested in astronomy.

* Schaaf, F, 1983, *The Wonders of the Sky*, Dover. Covers a range of phenomena from blue skies and rainbows through solar eclipses to planets and stars. Good first-hand accounts of many phenomena but lacks detailed explanations.

* Schaeffer, V.J. and Day, J.A., 1981, *A Field Guide to the Atmosphere*, Houghton Mifflin. Clouds are well covered, but sections on meteorological optics and ice and snow are patchy. Profusely illustrated with black and white photos. Appendix contains details of home experiments such as making permanent plastic casts of snowflakes.

Tricker, R.A.R., 1970, *Introduction to Meteorological Optics*, Mills & Boon. Not as wide ranging as Minnaert's book but the topics that it covers are dealt with in greater depth. Not suitable for someone without a background in physics since it is highly technical. Abbreviated in the Sources and notes as *Meteorological Optics*.

* Wood, E.A., 1975, *Science from your Airplane Window*, Dover. For the aeroplane passenger who wants explanations for the many phenomena that may be seen when flying: chapters on clouds and optics among others.

Magazines

There are many magazines that are of potential interest to anyone interested in natural phenomena. The ones below are probably the easiest to get hold of.

Astronomy: glossy, well-illustrated American monthly. Useful monthly information about astronomical events.

Astronomy Now: a British monthly aimed at the amateur astronomer.

Sky and Telescope: an excellent American monthly for amateur astronomers. Useful monthly information about astronomical events.

Weather: targeted at the serious amateur meteorologist. It has been published monthly since 1946 and is available on subscription from 'Weather', James Glaisher House, Grenville Place, Bracknell, Berks RG12 1BX, U.K.

Weatherwise: An excellent magazine for the amateur. It is published every two months. Available on subscription from: Weatherwise, 4000 Albermarle Street, NW, Washington, DC 20016.

Internet

The internet expands daily, and this list will be out of date almost as soon as I compile it. But it will serve as a starting point, particularly where atmospheric optics is concerned. Astronomical information is easier to track down on the internet because astronomy is a well established discipline. Astronomy is big science, with a huge amateur following, whereas atmospheric optics is very much a bit player.

The following sites deal with atmospheric optics:

Picture archive of atmospheric optics
http://www.meteoros.de/indexe.htm

Halos
http://www.sundog.clara.co.uk/halo/halosim.htm
http://www.student.oulu.fi/~jarkkoko/HALOS.HTM

Green flash
http://mintaka.sdsu.edu/GF/

Polarised light in nature
http://www.polarization.com/index.html

Crepuscular rays
http://www.ems.psu.edu/~demark/471/CrepuscularRays.html

Mirages
http://virtual.finland.fi/finfo/english/mirage.html

The following sites deal with astronomy

Solar System
http://www.seds.org/nineplanets/nineplanets/nineplanets.html

The Moon
http://seds.lpl.arizona.edu/billa/tnp/luna.html

Eclipses
http://www.uni-sw.gwdg.de/~bischoff/Sofi/sunearth.gsfc.nasa.gov/eclipse/eclipse.html

Sources and notes

Sources of references are listed alphabetically here for each section in the book. These include specific sources of quotations reproduced in the text, and more general sources of reference used during writing, and references that may be followed up by readers who wish to pursue a topic in more depth.

References to articles in journals give the author's surname followed by the year of publication, title of the article, name of the journal, its volume number and page numbers on which the article appears.

Abbreviations used for journals:

A.O.   Applied Optics
A.J.P.   American Journal of Physics
B.S.A.F.   Bulletin Societé Astronomique de France
J.B.A.A.   Journal of the British Astronomical Association
J.O.S.A.   Journal of the Optical Society of America
M.M.   Meteorological Magazine
M.O.   Marine Observer
M.W.R.   Monthly Weather Review
P.T.   Physics Teacher
Q.J.R.A.S.   Quarterly Journal of the Royal Astronomical Society
Q.J.R.M.S.   Quarterly Journal of the Royal Meteorological Society
Sc.Am.   Scientific American
S. and T.   Sky and Telescope

As well as sources, there are some supplementary notes on the text.

Chapter 1: Daylight

1.1 The colour of the daytime sky

Bohren, C.F, 1987, *Clouds in a Glass of Beer*, John Wiley, pp. 160–3.
Können, G.P., 1985, *Polarized Light in Nature*, Cambridge University Press, pp. 29–31.
Lynch, D.K. and Livingston, W., 1995, *Colour and Light in Nature*, Cambridge University Press, chapter 2.

Minnaert, M., 1954, *Light and Colour in the Open Air*, Dover, pp. 238–54.
Tricker, R.A.R., 1970, *Introduction to Meteorological Optics*, Mills and Boon, chapter 9.

1.2 Why is the sky blue?

Bohren, C.F. and Fraser, A.B., 1985, 'Colours of the Sky', *P.T.*, 23, pp. 267–72.
da Vinci, Leonardo, 1980, *The Notebooks of Leonardo da Vinci*, Oxford University Press, see 'The Four Elements' pp. 37–9 for Leonardo's ideas on the colour of the sky.
Newton, Issac, 1952, *Optics*, Dover. See Book 2, part 3, proposition 7 for Newton's thoughts on the colour of the sky.
De Saussure, H. B., 1788–1789, *Description d'un cyanometre ou d'un appareil destiné a mesurer la couleur bleue du ciel*, Memoires. de l'Acad. Roy. des Sciences de Turin, IV. This is an account of De Saussure's researches on sky colours.

1.3 Airlight

Greenler, R., 1980, *Rainbows, Halos and Glories*, Cambridge University Press, pp. 133–4.
McCartney, E.J., 1976, *Optics of the Atmosphere, Scattering by Molecules and Particles*, Wiley, pp. 20–44.
Went, F.W., 1960, 'Blue Hazes in the Atmosphere', *Nature*, 187, pp. 641–3.
Minnaert, M., *Light and Colour*, pp. 235–7.

1.4 Aerial perspective

Bohren, C.F., *Clouds*, chapter 16.
Cornish, V., 1926, 'Harmonies of Tone and Colour in Scenery Determined by Light and Atmosphere', *Geographical Journal*, 26, pp. 507–28.
Middleton, W.E.K., 1952, *Vision Through the Atmosphere*, University of Toronto Press.
Darwin, C., 1959, *Voyage of the Beagle*, J.M. Dent, p. 312.

1.5 How far can you see?

Corfidi, S.F., 1993, 'Those Hazy Days of Spring and Summer', *Weatherwise*, June–July, pp. 13–17.

Bohren, C.F. and Fraser, A.B., 1986, 'At What Altitude Does the Horizon Cease to be Visible?', *A.J.P.*, 54, pp. 222–7.

1.6 Polarised light

Bohren, C.F., 1991, *What Light Through Yonder Window Breaks*, John Wiley, chapter 4.
Können, G.P., *Polarized Light*.
Lynch, D.K. and Livingston, W., *Colour and Light*, section 2.4.
Shurcliff, W.A. and Ballard, S.S., 1964, *Polarized Light*, Van Nostrand.

1.7 Polarised light from the sky

Use a camera polarising filter. These are relatively inexpensive and widely available and, being circular, are easily rolled between the fingers when testing for polarised light.

1.8 Polarised light due to reflections

The Brewster angle for water is 53°.

1.9 Haidinger's brush

Shurcliff, W.A. and Ballard, S.S., *Polarized Light*, pp. 95–8.

Chapter 2: Shadows

2.1 No light without shadow

Kepler, J., 1604, *Ad Vitellionem paralipomena, quibus Astronomiae pars optica traditur*. The first English translation of the complete text of Kepler's seminal work on optics, by William H. Donahue, published by Green Lion Press, 2000. Contains Kepler's explanation of shadows.

2.2 Solar shadows

The Moon's diameter is 3476 km and so the maximum reach of its umbral shadow is $3476 \times 120 = 417\,120$ km. Moon's average orbital radius 384 500 km. Hence the maximum distance at which the Moon's umbra can be formed is almost the same as the Moon's average orbital radius. What an extraordinary coincidence.

Lynch, D.K. and Livingston, W., *Colour and Light*, chapter 1.

2.3 Shadows formed by point sources

Rudaux, L. and de Vaucouleurs, G., 1966, *Larousse Encyclopedia of Astronomy*, Paul Hamlyn. There is a photo of shadow cast by Venus (p. 188). Diffraction rings are visible around the edge of the shadow.
Kiernan, N.S., 1985, 'Shadows Cast by Venus', *J.B.A.A.*, 95, p. 272.
Parish, P.W., 1986, 'Shadows Cast by Jupiter and Sirius', *J.B.A.A.*, 96, p. 10.

2.4 Mach bands

Enright, J.T., 1994, 'Mach Bands and Airplane Shadows Cast On Dry Terrain', *A.O.*, 33, pp. 4723–26.
Ratliffe, F., 1972, 'Contour and Contrast', *Sc.Am.*, 226, pp. 90–101 (June).
Sekuler, R. and Blake, R., 1994, *Perception*, McGraw-Hill, 3rd edn, pp. 74–6.

2.5 Coloured shadows

High-pressure sodium lights have a broader spectrum than low-pressure ones and so don't produce such obvious coloured shadows.

Churma, M.E., 1994, 'Blue Shadows, Physical, Physiological and Psychological Causes', *A.O.*, 33, pp. 4719–22.
Falk, D., Brill, D. and Stork, D., 1986, *Seeing the Light*, Harper and Row. Particularly good on the optics and physiology of vision. See pp. 280–3 on simultaneous colour contrast
Goethe, J.W. von, 1970, *Theory of Colours*, MIT Press. The record of the fruits of Goethe's attempts to gather evidence against the Newtonian account of light and colour. See section 6 for coloured shadows.
Mollon, J., 1995, 'Seeing Colour' in Lamb, T. and Bourriau, J. (eds.), 1995, *Colour: Art and Science*, Cambridge University Press, pp. 147–9.
Sekuler, R. and Blake, R.,1994, *Perception*, McGraw-Hill (3rd edn); chapter 6 deals with colour perception.

2.6 The *heiligenschein*

Evershed, J., 1913, 'Luminous Halos Surrounding Shadows of Heads', *Nature*, 90, p. 592.

Floor, C., 1983, 'The Heiligenschein', *Weather*, 38, pp. 41–4.

Fraser, A.B., 1994, 'The Sylvashine, Retroreflection from Dew-covered Trees', *A.O.*, 33, pp. 4539–747.

Greenler, R., *Rainbows*, pp. 146–9.

Lynch, D.K. and Livingston, W., *Colour and Light*, section 4.12.

Mattsson, J.O. and Cavallin, C., 1972, 'Retroreflection of Light from Drop Covered Surfaces', *Oikos*, 23, pp. 285–94.

Minnaert, M., *Light and Colour*, pp. 254–7.

Preston, J.S., 1969, 'Increased Luminance in the Direction of Retroreflection', *Nature*, 224, p. 1102.

Tricker, R.A.R., *Meteorological Optics*, pp. 24–41.

2.7 Shadows on water

Jacobs, S.F., 1953, 'Self-Centred Shadow', *A.J.P.*, 21, p. 234.

Walker, J., 1987, 'Reflections From a Water Surface Display: Some Curious Properties', *Sc.Am.*, 256, pp. 108–114 (January).

Walker, J., 1988, 'Shadows Cast On The Bottom Of a Pool Are Not Like Other Shadows. Why?', *Sc.Am.*, 259, pp. 86–9 (July).

2.8 Shadows formed by clouds

Lynch, D.K., 1987, 'The Optics of Sunbeams', *J.O.S.A.*, 4, pp. 609–11.

Lynch, D.K. and Livingston, W., *Colour and Light*, section 1.9.

Chapter 3: Mirages

3.1 Atmospheric refraction

Fraser, A.B. and Mach, W.H., 1980, 'Mirages' in *Atmospheric Phenomena*, W.H. Freeman, pp. 81–9.

Greenler, R., *Rainbows*, chapter 7.

Können, G.P., *Polarized Light*, p. 78.

Lynch, D.K. and Livingston, W., *Colour and Light*, sections 2.23–2.26.

Minnaert, M., *Light and Colour*, pp. 43–58.

Tape, W., 1985, 'The Topology of Mirages', *Sc.Am.*, 252, pp. 100–8 (June).

Tricker, R.A.R., *Meteorological Optics*, pp. 19–20.

3.2 Inferior mirages

Ashmore, S.E., 1955, 'A North Wales Road Mirage', *Weather*, 10, pp. 336-42.
Bohren, C.F., *What Light*, chapter 6.
Fraser, A.B., 1975, 'Theological Optics', *A.O.*, 14, pp. 92–3.
Monge G., 1902, 'Sur le phenomene d'optique, connu sous le nom de mirage', in *Neudrucke von Schriften und Karten über Meteorologie und Erdmagnetism*, vol. 14, ed. G. Hellman, Berlin, pp. 74–81.
Vedy, L.G., 1928, 'Sand Mirages', *M.M.*, 63, pp. 249–55.

3.3 Superior mirages

Ives, R.L., 1968, 'The mirages of La Encantada', *Weather*, 23, pp. 55–60.
Lown, K.R., 1987, 'Mirage over the Thames Estuary', *Weather*, 42, p. 393.
Scoresby, W., 1902, 'Descriptions of some remarkable Atmospheric Reflections and Refractions, observed in the Greenland Sea', in *Neudrucke von Schriften und Karten über Meteorologie und Erdmagnetism*, vol. 14, ed. G. Hellman, Berlin, pp. 82–6.
Shackelton, E., 1916, *South!*, Heinemann.
Tyrrell, J.B. (ed.), 1916, *David Thompson's Narrative of his Explorations in Western America 1784–1812*, Champlain Society, Canada, p. 120.
Vince, Rev. S., 1799, 'Observations on an unusual Horizontal Refraction of the Air', *Philosophical Transactions of the Royal Society*, 89, pp. 13–25.
Vollprecht, R.V., 1947, 'The "Cold Mirage" in Western Australia', *Weather*, 2, pp. 174–8.

3.4 Lake monsters

Lehn, W.H., 1979, 'Atmospheric Refraction and Lake Monsters', *Science*, 205, pp. 183–5.
Lehn, W.H., 1981, 'The Norse Merman As An Optical Phenomenon', *Nature*, 289, pp. 362–6.

3.5 Looming and unusual visual range

Hobbs, W.H., 1937, 'Conditions of Exceptional Visibility within High Latitudes, Particularly as a result of Superior Mirage', *Annals of the American Association of Geographers*, Dec., pp. 229–240.
Latham, W., 1798, 'On A Singular Instance Of Atmospherical Refraction', *Philosophical Transactions of the Royal Society*, 88, p. 357.

Sawatzky H.L. and Lehn, W.H., 1976, 'The Arctic Mirage And The Early North
    Atlantic', *Science*, 192, pp. 1300–5.
Southey, D.J., 1992, 'Mirage over Hastings', *Weather*, 47, p. 320.
Thomas, F.G., 1993, 'Mirage over Hastings – 5 Aug 1987', *Weather*, 48, p. 30.

Chapter 4: Sunset and sunrise

4.2 Twilight

Lynch, D.K. and Livingston, W., *Colour and Light*, sections 2.9–2.16.
Meinel, A. and Meinel M., *Sunsets*, chapter 4.
Cornish, V., 1926, 'Harmonies of Tone and Colour in Scenery Determined by Light
    and Atmosphere', *Geographical Journal*, 26, pp. 506–28.

4.3 Clouds at sunset

Atkins, W.R.G., 1945, 'Apparent Clearing of the Sky at Dusk', *Nature*, 155, p. 110.

4.4 The purple light

Peterson, R.E., 1992, 'Atmospheric Effects Caused By Mt Pinatubo Eruptions
    Observed In Texas', *Weather*, 47, pp. 164–6.

4.5 Crepuscular rays

Goin, P., 1968, 'Unusual Twilight Phenomena over Ethiopia', *Weather*, 23, pp.
    70–1.
Minnaert, M., *Light and Colour*, pp. 275–7.
Monteith, J.L., 1986, 'Crepuscular Rays Formed By The Western Ghats', *Weather*,
    41, pp. 292–4.

4.6 Mountain shadows

Livingston, W. and Lynch, D., 1979, 'Mountain Shadow Phenomena', *A.O.*, 18, pp.
    265–9.
Meinel, A. and Meinel M., *Sunsets*, pp. 31–2.

4.7 Abnormal twilights

Austin, J., 1983, 'Krakatoa Sunsets', *Weather*, 38, pp. 226–31.
Meinel, A and Meinel M, *Sunsets*, chapter 5.

4.8 The Sun at the horizon

Floor, C., 1982, 'The Setting Sun', *Physics Education*, 17, pp. 174–7.

4.9 Green flashes

Bohren, C.F., *Clouds*, chapter 13.
Botley, C.M., 1971, 'The Green Ray', *Weather*, 26, pp. 354–7.
Fraser, A.B., 1975, 'The Green Flash And Clear Air Turbulence', *Atmosphere*, 13, pp. 1–10.
Minnaert, M., *Light and Colour*, pp. 58–63.
O'Connell, D.J.K., 1968, *The Green Flash and other Low Sun Phenomena*, North Holland.
O'Connell, D.J.K., 1980, 'The Green Flash', in *Atmospheric Phenomena*, ed. D. Lynch, W.H. Freeman.
Shaw, G. E., 1973, 'Observations And Theoretical Reconstructions Of The Green Flash', *Pure and Applied Geophysics*, 102, pp. 223–35.

4.10 The Purkinje effect

Pirenne, M.H., *Vision and the Eye*, Chapman and Hall, pp. 38–9.
Sekuler, R. and Blake, R., *Perception*, pp. 94–5.

Chapter 5: Rainbows

5.1 Unweaving the rainbow

Boyer, C.B., 1987, *The Rainbow*, Macmillan. An excellent history of views on the nature of the rainbow from the earliest times to the present.
Gage, J., 1993, *Colour and Culture*, Thames and Hudson. See chapter 6 for a comprehensive account of rainbows in European art.
Lee, R. L. and Fraser A.B., 2001, *The Rainbow Bridge: Rainbows in Art, Myth, and Science*, Penn State University Press.

5.2 How to recognise a rainbow

Bohren, C.F., *Clouds*, chapters 21 and 22.
Gedzelman, S.D., 1980, 'Visibility of Halos and Rainbows', *A.O.*, 19, pp. 3068–74.
Greenler, R., *Rainbows*, chapter 1.
Lee, R. and Fraser, A.B., 1990, 'The Light at the End of the Rainbow', *New Scientist* (1 Sept), pp. 40–4.
Lynch, D.K. and Livingston, W., *Colour and Light*, sections 4.1–4.11.
Minnaert, M., *Light and Colour*, chapter 10.
Tricker, R.A.R., *Meteorological Optics*, chapters 3 and 6.

5.4 Supernumerary bows.

Brewster, D., 1829, 'Note', *Brewster's Journal of Science*, 10, p. 163.
Langwith, Rev Dr. (Benjamin), 1722, 'Concerning the appearance of several arches of colour contiguous to the inner edge of the common rainbow. First Letter, Petworth, Jan 12', *Philosophical Transactions of the Royal Society*, vol 32.
Walker, D., 1950, 'A Rainbow and Supernumeraries with Graduated Separations', *Weather*, 5, p. 324.

5.5 Circular rainbows

'Marine Observer's Log' in *M.O.*, 31, 191, 1961.

5.6 Rainbows at sunset and sunrise

Harries, H., 1922, 'Letter', *M.M.*, 57, p. 246.
Palmer, F., 1945, 'Unusual Rainbows', *A.J.P.*, 13, pp. 203–4.
Thompson, A.H., 1974, 'Water bows, white bows and red bows', *Weather*, 29, pp. 178–84.

5.7 Lunar rainbows

There is a photograph of a lunar rainbow by William Sager showing the expected colours in *S.& T.*, 59, 177, 1980.
Brook, C.L., 1900, 'Double Lunar Rainbow', *Symons's Meteorological Magazine*, 35, p. 138.
Humphreys, W.J., 1938, 'Why we Seldom see Lunar Bows', *Science*, 88, pp. 496–8.
Wentworth, C.K., 1938, 'Frequency of Lunar Rainbows', *Science*, 88, p. 498.

5.8 Reflection rainbows

Celsius, A., 1743, 'Sur un Arc-en-ciel extraordinaire vu en Dalecarlie' *L'Histoire de l'Academie de Science*, 35.
Halley, E., 1698, Untitled letter, *Philosophical Transactions of the Royal Society*, (May edition).
Frost, P., 1889/90, Untitled letter, *Nature*, 41, p. 316.

5.9 Reflected rainbows

Dawson, G., 1874, 'Rainbow and its reflection', *Nature*, 9, p. 322.

5.10 Spray bows

A photograph on p. 54 of G.P.Können's *Polarized light in Nature* shows a spray bow formed in sea spray at the same time as a rainbow. It shows clearly the difference between the angular radius of these bows.

Martin, E.A., 1911, Untitled letter, *Nature*, 87, p. 450.

5.11 Fog bows

There is a photograph by Dianne March of a fog bow that looks like a low arch on p. 19 of *Weatherwise* for Aug/Sept. 1993.

Books, C.F., 1925, 'Coronas and Iridescent Clouds', *M.W.R.*, 53, p. 54.
Lynch, D.K. and Schwartz, P., 1991, 'Rainbows and Fogbows', *A.O.*, 30, pp. 3415–20.
McDonald, J.E., 1962, 'A Gigantic Horizontal Rainbow', *Weather*, 17, pp. 243–5.
Palmer, F., 1945, 'Unusual Rainbows', *A.J.P.*, 13, pp. 203–4.

5.12 Rainbow wheels

Thompson, S. P., 1878, 'On certain phenomena accompanying rainbows', *Philosophical Magazine*, 6, pp. 272–4.
Mark Twain, *Roughing It*, chapter 71.
There is a rainbow like the one described by S.P. Thompson in a small painting titled *London from Hampstead Heath, with a double rainbow* by the English painter John Constable.

5.13 Horizontal rainbows

Lord Dulverton, Untitled letter, *The Times*, 15 November, 1937.

Humphreys, W.J., 1929, 'The Horizontal Rainbow', *Journal of the Franklin Institute*, 30, pp. 661–4.

Juday, C., 1916, 'Horizontal Rainbows On Lake Mendota', *M.W.R.*, 44, pp. 65–7.

Monteith, J.L., 1954, 'Refraction and the Spider', *Weather*, 9, pp. 140–1.

Mattsson, J.A., Nordbeck, S. and Rystedt, B., 1971, 'Dewbows and Fogbows in Divergent Light', *Lund Studies in Geography*, Number 11.

5.14 Seachlight rainbows

Floor, C. 1986, 'Rainbows and Haloes in Lighthouse Beams', *Weather*, pp. 203–8.

Harsch, J. and J.D. Walker, 1975, 'Double Rainbow and Dark Band in a Searchlight Beam', *A.J.P.*, 43, pp. 453–5.

5.15 Eclipse rainbows

Maunder, E.W., 1901, 'The Comet and the Eclipse', *The Observatory*, 24, pp. 372–6.

5.16 Anomalous rainbows

'An account . . . of two rainbows, unusually posited, seen in . . . France . . .', *Philosophical Transactions of the Royal Society*, vol. 1, 1665.

Betlem, H. and Zwart, B., 1975, 'Nerkwaardig regenboogverschijnesel (Remarkable rainbow phenomenon)', *Zenit*, 2, p. 402.

Corliss, W.R. (ed.), 1983, *Unusual Natural Phenomena*, Doubleday Anchor. Subtitled 'Eyewitness accounts of nature's greatest mysteries', this book is part of a series in which Corliss has collected together first-hand accounts of a broad range of unusual phenomena. The book gives many examples of anomalous rainbows. No explanations are offered. Other books in this series cover astronomical, geological and meteorological phenomena. These are available from the author, Roger Corliss, Sourcebook Project, P.O. Box 107, Glen Arm, MD 21057, U.S.A.

Glaisher, J., 1861–63, Untitled letter dated 6 November 1861, *Proceedings of the British Meteorological Society*, vol. 1.

Hannay, J.B., 1879, Untitled letter to Editor, *Nature*, 21, p. 56.

Laine, V.J., 1909, 'Effect of Lightning on a Rainbow', *Physikalische Zeitschrift*, 10, p. 965.

Poey, A., 1863, 'Rainbows and Supernumerary Bows Observed in Havana', *Compte Rendus*, 57, p. 109.

Voltz, F.E., 1960, 'Some Aspects of Optics of the Rainbow and the Physics of Rain', *Physics of Precipitation*, American Geophysical Union, pp 230–86,

Wood, R.W., 1906, 'Review of Perntner's Meteorological Optics', *M.W.R.*, 34, p. 356.

5.17 Explaining the rainbow

Fraser, A.B., 1983, 'Chasing Rainbows, Numerous Supernumeraries are Super', *Weatherwise*, 36, pp. 280–9.

Tricker, R.A.R., *Meteorological Optics*, chapter 3.

Walker, J.D., 1976, 'Multiple Rainbows from Single Drops of Water and Other Liquids', *A.J.P.*, 44, pp. 421–33.

5.18 Tertiary bows

Pegley, D.E., 1986, 'A Tertiary Rainbow', *Weather*, 41, p. 401.

5.19 Polarised rainbows

Können, G.P., *Polarized Light*, pp. 46–56.

5.20 Are rainbows real?

When my son was a little boy he once suggested that the increased brightness that is often seen at the foot of a rainbow led people to believe that it must be due to a pot of gold glinting in the sunlight.

5.21 Notes for rainbows observers

Share your observations of unusual rainbows by writing to *Weather*.

Chapter 6: Coronae and glories

6.1 Coronae

Bohren, C.F., *Clouds*, chapter 17.

Books, C.F., 1925, 'Coronas and Iridescent Clouds', *M.W.R.*, 53, pp. 49–58.

Greenler, R. *Rainbows*, chapter 6.

Lynch, D.K. and Livingston, W., *Colour and Light*, section 4.13.

Minnaert, M., *Light and Colour*, pp. 214–23.

Tricker, R.A.R., *Meteorological Optics*, chapter 5.

6.4 Glories

Black, D.M., 1954, 'The Brocken Spectre of the Desert View Watch Tower, Grand Canyon, Arizona', *Science*, 119, pp. 164–5.

Bouguer, P., 1902, 'Brockenspenst', in *Neudrucke von Schriften und Karten über Meteorologie und Erdmagnetism*, vol. 14, ed. G. Hellman, Berlin, pp. 52–5.

Bryant, H.C. and Jarmie, N., 1980, 'The Glory', in *Atmospheric Phenomena*, ed. D.K. Lynch, W.H. Freeman.

Greenler, R., *Rainbows*, pp. 143–6.

Können, G.P., *Polarized Light*, pp. 69–72.

Lynch, D.K. and Livingston, W., *Colour and Light*, section 4.15.

Tricker, R.A.R., *Meteorological Optics*, chapter 7.

Minnaert, M., *Light and Colour*, pp. 224–6 and 257–61.

Windass, C., 1948, 'A Glory of Sea Fog', *Weather*, 3, p. 18.

Wood, E.A., *Science from your Airplane Window*, pp. 68–71.

Wright, C.J., 1980, 'The Spectre of Science. The Study of Optical Phenomena and the Romantic Imagination', *Journal of the Warburg and Courtauld Institute*, 43, pp. 186–200.

6.5 Notes for observers of coronae and glories

Share your observations of unusual glories by writing to *Weather*.

Chapter 7: Atmospheric halos

7.1 Ice halos

Brain, J.P., 1972, Halo Phenomena – An Investigation, *Weather*, 27, pp. 409–10.

Goldie, E.C.W., 1971, 'A graphical Guide to Haloes', *Weather*, 26, pp. 391–4.

Greenler, R., *Rainbows*, chapters 2, 3 and 4.

Können, G.P., *Polarized Light*, pp. 56–69.

Lynch, D.K., 1980, 'Atmospheric Halos', in *Atmospheric Phenomena*, ed. D.K. Lynch, W.H. Freeman.

Lynch, D.K. and Livingston, W., *Colour and Light*, chapter 5.

Minnaert, M., *Light and Colour*, pp. 190–208.

Tape, W., 1994, *Atmospheric Halos*, American Geophysical Union.
  Comprehensive, non-technical coverage of halos relating them to the crystals
  that cause them. Wonderful photographs.
Tricker, R.A.R., *Meteorological Optics*, chapter 4.

7.2 22° halo

Gedzelman, S.D., 1980, 'Visibility of Halos and Rainbows', *A.O.*, 19, pp. 3068–74.
Tape, W., *Atmospheric Halos*, chapter 4.

7.3 Upper tangent arc

Tape, W., *Atmospheric Halos*, pp. 13–17.

7.4 46° halo

Greenler, R., *Rainbows*, pp. 47–50.
Tape, W., *Atmospheric Halos*, chapter 4.

7.5 Rare halos

Goldie, E.C.W., Meaden, G.F. and White, R., 1976, 'The Concentric Halo Display of
  14 April 1974', *Weather*, 31, p. 304.
Greenler, R., *Rainbows*, pp. 61–4.
Tape, W., *Atmospheric Halos*, chapters 10 and 11.
Turner, F.M. and Radke, L.F., 1975, 'Rare Observation of the 8° Halo', *Weather*, 30,
  pp. 150–5.

7.6 Parhelia

Brain, J.P., 1972, 'Halo Phenomena – An Investigation', *Weather*, 27, pp. 409–10.

7.7 Circumzenithal and circumhorizontal arcs

Greenler, R., *Rainbows*, pp 50–2.
Tape, W., *Atmospheric Halos*, chapter 1.

7.8 The parhelic circle

There is a photograph by Grant Goodge of a parhelic circle with an angular radius of 10° on p. 9 of *Weatherwise*, June–July, 1992.

7.10 Subsuns

Mattsson, J.O., 1973, 'Subsun and Light Pillars of Street Lamps', *Weather*, 28, pp. 66–8.

7.11 Notes for ice halo observers

Report unusual halos by writing to *Weather* or sending an e-mail to one of the various sites dedicated to atmospheric optics.

Chapter 8: The night sky

8.1 A brief history of the sky

There are many histories of astronomy. I recommend the following.
Pannekoek, A., 1989, *A History of Astronomy*, Dover.
Hoskins, M. (ed.), 1999, *The Cambridge Concise History of Astronomy*, Cambridge University Press.
Kuhn, T. S., 1995, *The Copernican Revolution*, Harvard University Press.

8.2 Naked-eye astronomy

Davidson, N. 1985, *Astronomy and the Imagination*, RKP. This deals with earth-centred, naked-eye astronomy. Unfortunately, the author takes an unfashionably metaphysical view of astronomical phenomena, and this led to some unfavourable reviews in the scientific press. Nevertheless, it contains far more information about how the sky actually works than most books on astronomy.
Muirden, J., 1963, *Astronomy with Binoculars*, Faber. Many useful activities for the naked-eye observer.
Ottewell, G., 1989, *The Astronomical Companion*, Astronomical Workshop, Furman University, Greenville, S.C. 29613, U.S.A. The most imaginative and exciting visual guide to the night sky that anyone has ever written. Several other publications dealing with astronomical matters by the same author are all available from the same address.

Schaaf, F., 1990, *Seeing the Sky*, J. Wiley. Subtitled '100 Projects, Activities and Explorations in Astronomy'. Invaluable source of naked-eye activities for amateur observers. Covers meteorological optics as well as the night sky.

Norton, A.P., ed. Satterthwaite, G.P., 1986, *Norton's Star Atlas*, Longman. Detailed sky maps, and a great deal of astronomical information besides.

8.3 The celestial sphere

Strictly speaking, only stars well away from the celestial pole rise and set at the horizon. Those closer to the pole, so-called circumpolar stars, are always above the horizon and move around the pole. This merely reinforces the illusion that the heavens are rotating. Polaris, of course, is effectively stationary because it lies so close to the north pole that its motion around it in the course of a night is not detectable by the naked eye.

8.4 The ecliptic

Computer simulations of the night sky are invaluable aids for the stargazer. However, there are so many simulations that I hesitate to recommend one in particular. You will find reviews of such computer software in most astronomy magazines. A planisphere, a star map printed on a circular disc of stiff plastic, makes a good alternative to a computer simulation. It is portable and can be made small enough to fit in your pocket.

8.6 Why is the sky dark at night?

Harrison, E., 1987, *Darkness at Night*, Harvard University Press. This is probably the most comprehensive book on the history of this fascinating topic.

8.7 The sky beyond the equator

Davidson, N., 1993, *Sky Phenomena*, Floris Books, chapter 9: 'The Southern Hemisphere Sky'.

Chapter 9: The Moon

9.2 The Earth–Moon system

Although ancient Greek astronomers established that the Sun was several times larger than the Earth, their estimates of the Sun's diameter were at least an order of magnitude too small.

Recent (1987) measurements suggest that Pluto's moon, Charon, is the largest moon in the Solar System relative to the parent planet. The largest moon in absolute terms is Ganymede, diameter 5260 km, which orbits Jupiter. Titan, which orbits Saturn, is the second largest moon and has a diameter of 5150 km. Some books make Titan larger than Ganymede, but do so by adding the depth of Titan's atmosphere to its diameter. Mercury, a planet in its own right, has a diameter of 4880 km.

Eclipses of the Sun occur elsewhere in the Solar System: For example, Io has an apparent diameter that is some 5 times that of the Sun seen from Jupiter. Io can therefore eclipse the Sun, and does so more often than our Moon because it orbits more quickly (in 1.77 days). Jupiter's disc is about 10 times that of the Earth, and hence more likely to intercept Io's shadow.

Rothery, D.A., 1999, *Satellites of the Outer Solar System*, Oxford University Press. An excellent, up to date account of the origin of natural satellites, including the Moon.

9.3 Looking at the Moon without a telescope

Flammarion, C., 1900, 'Dessins de la Lune Vue a l'Oeil nu', *B.S.A.F.*, 14, pp. 45–50, 93–8, 140–5, 183–8, 227–33, 275–83, 498–506.

Plutarch, 1957, 'De facie in orbe lunae' *Plutarch Moralia*, volume XII, Loeb Classical Library, Translated by H. Cherniss.

Sheehan, W., 1988, *Planets and Perception*, University of Arizona Press. See chapter 2 for pre-telescopic views of the Moon.

9.4 The lunar surface

Montgomery, S.L., 1999, *The Moon and Western Imagination*, University of Arizona Press, pp. 86–96.

The painter Jan van Eyck appears to be the first person to have drawn a naturalistic representation of the Moon. It can be seen in one of his paintings, *The Crucifixion* (painted 1420–25), in which there is a tiny gibbous Moon that shows recognisable maria. Leonardo da Vinci also drew of the Moon's surface. There are two small sketches in his notebooks for the years 1505 to 1508.

Galileo Galilei, 1957, *Discoveries and Opinions of Galileo*, translated with an introduction and notes by Stillman Drake, Doubleday Anchor. Includes a new translation of *The Starry Messenger*. Galileo's account of what are in most cases original observations of new phenomena are both fascinating and stirring.

Wood, E.A., 1975, *Science from your Airplane Window*, Dover. See chapter 2 for a discussion of the appearance of water from a height.

Whitaker, E.A., 1989, 'Selenography in the seventeenth century' in *The General History of Astronomy*, vol. 2: *Planetary astronomy from the Renaissance to the Rise of Astrophysics*. Part A 'Tycho Brahe to Newton', ed. R.Taton and C. Wilson, Cambridge University Press, pp. 119–43. This paper has copies of both Gilbert's and da Vinci's naked-eye sketches of the Moon.

Brown, G.C., Hawkesworth C.J. and Wilson, R.C.L. (eds.), 1992, *The Origin of the Earth*, Cambridge University Press. For an exposition of current views of the formation of the Earth–Moon system see chapter 2: 'Understanding the Earth' by S.R.Taylor.

9.6 The Moon in daylight

Danjon, A., 1932, 'Jeunes et vielles lunes', *B.S.A.F.*, 46, pp. 57–66.

Danjon, A., 1936, 'Le croissant lunaire', *B.S.A.F.*, 50, pp. 57–65.

9.7 How bright is the Moon?

Bohren, C.F., *What Light*, chapter 15.

Minnaert, M., *Light and Colour*, pp. 80–1. Minnaert claims that as Herschel arrived at Cape Town by sea, in January 1834, he saw an almost-full Moon rising over Table Mountain and that this prompted him to make his ingenious estimate of the Moon's brightness by comparing it to that of Table Mountain there and then. However, this can't be right, because in his diaries, Herschel records observations of a lunar eclipse on 28 December 1833 when he was still at sea. This was two weeks before he arrived at Cape Town on 15 January 1834, when, of course, the Moon would have been at or near the start of a new lunation. In any case, Herschel outlines his method of comparing the Moon's surface with that of the Earth in a footnote on page 272 of the 1881 edition of his book *Outlines of Astronomy*.

9.8 Earthshine

Plassmann, J. P., 1924, 'Studien uber das aschgraue Mondlicht', *Die Himmelswelt*, 34, pp. 95–103.

Plassmann, J. P., 1928, 'Aus der morgenländischen Sternkunde und Sterndeuterei', *Die Himmelswelt*, 38, pp. 20–25.

9.9 Moonlight

Smith, G., Vingrys, A.J., Maddocks, J.D. and Hely, C.P., 1994, 'Color Recognition and Discrimination Under Full-Moon Light', *A.O.*, 33, pp. 4741–8.

9.10 The Moon illusion

Hershenshon, M. (ed.), 1989, *The Moon Illusion*, Lawrence Erlbaum Associates. This is a definitive collection of the several schools of thought on the subject, together with a comprehensive historical survey of the phenomenon. Huge bibliography.

Rock, I., 1984, *Perception*, Scientific American Library. See pp. 26–30 for a clear treatment of a widely held explanation of the Moon illusion.

9.11 A Blue Moon

Bohren, C.F., *Clouds*, chapter 12.

Botley, C.M., 1936, 'Blue Moon', *Q.J.R.M.S.*, 62, p. 500.

Bull, G.A., 1951, 'Blue Sun and Moon', *M.O.*, 21, pp. 153–67.

Meinel, A. and Meinel M., *Sunsets*, pp. 83–5.

Peacock, J.M., 1990, 'Blue Sun once in a Blue Moon', *Weather*, 45, p. 158.

9.16 Libration

Meketa, J.E., 1987, 'Viewing Libration With the Naked Eye', *S.&T.*, 74 (July), p. 63.

Muirden, J., 1963, *Astronomy with Binoculars*, Faber, p. 33.

Have a look at the section on the Moon in the 'Nine Planets' website for an animated sequence that shows how the Moon wobbles about as it orbits the Earth.

9.17 Lunar puzzles

Brewer, S.G., 1988, *Do-it-yourself Astronomy*, Edinburgh University Press, pp. 104–6.

9.18 The Moon's phases

Alastair McBeath, an experienced observer, has told me that in 18 years of observing he has yet to see a young Moon less than 24 hours old.

Barlow Pepin, M., 1996, 'In Quest of the Youngest Moon', *S.&T.*, 92, p. 104.

Cherrington, E., 1969, *Exploring the Moon Through Binoculars*, Dover.

di Ciccio, D., 1989, 'Breaking the New-Moon Record', *S.&T.*, 78, p. 322.

Danjon, A., 1932, 'Jeunes et vielles lunes', *B.S.A.F.*, 46, pp. 57–66.

Danjon, A., 1936, 'Le croissant lunaire', *B.S.A.F.*, 50, pp. 57–65.

Hopkins, B.J., 1883, 'Abnormal Appearance of the Lunar Crescent', *The Observatory*, 6, p. 247. Hopkins noted the invisibility of the Moon's horns several decades before Danjon, but his observation does not appear to have stirred any interest because there is no further correspondence on the subject in the journal.

Gavin, M., 1984, 'The Young Crescent Moon at Apogee, 30.9 Hours Old', *J.B.A.A.*, 94, p. 187. Gavin reports that he saw the young Moon on 2 April 1984 when it was 14° west of the Sun. 'A pair of binoculars was necessary . . . in a bright but crystal clear western sky . . . the faintest sliver of the crescent moon extending about 120° at most . . . the moon was only 5° above the horizon.'

Chapter 10: Eclipses

10.2 Solar eclipses

Littmann, M. and Willcox, K., 1991, *Totality: Eclipses of the Sun*, University of Hawaii Press. Information and advice on all aspects of solar eclipses for the amateur.

Meus, J., 1995, 'Solar Eclipse Diary, 1995–2005', *S.&T.*, 89 (Feb), pp. 29–32. Details, including a map, of when and where eclipses of the Sun will occur during these years.

Zirker, J.B., 1984, *Total Eclipses of the Sun*, Van Nostrand. Thorough, non-technical account of solar eclipses.

Codona, J.L., 1991, 'The Enigma Of Shadow Bands', *S.&T.*, 81 (May), pp. 482–7.

Chapter 11: Planets

11.1 The solar system

Taylor, S.R., 1998, *Destiny or Chance: Our Solar System and its Place in The Cosmos*, Cambridge University Press. This is a comprehensive account of current theories concerning the formation of the Solar System.

Grosser, M., 1979, *The Discovery of Neptune*, Dover. See chapters 1 and 2 on the discovery of Uranus.

Evidence that there really are planets orbiting other stars was first obtained during the last decade of the twentieth century. Since these planets can't be seen, astronomers rely upon the fact that a massive planet will cause the star it orbits to wobble slightly. This wobble is detectable.

11.2 How to tell a planet from a star

Edberg, S.J. and Levy, D.H., 1994, *Observing Comets, Asteroids, Meteors, and the Zodiacal Light*, Cambridge University Press. See chapter 4 on asteroids.

Schaaf, F., 1990, *Seeing the Sky*, J. Wiley. See section 20 on 'Trying to see the moons of Jupiter with the naked eye'.

11.4 Where to look for an inferior planet

It is sometimes possible to see Venus at midnight at high latitudes during summer when the sun itself is just below the horizon. If Venus is above the ecliptic, it may just be circumpolar at high latitudes and hence above the northern horizon.

11.6 Venus

Bobrovnikoff, N., 1990, *Astronomy Before the Telescope*, vol 2, *The Solar System*, Pachart Publishing House. See remarks on the naked-eye visibility of planets, p. 48.

Ellis, E.L., 1995, 'Naked-eye Observations of Venus in Daylight', *J.B.A.A.*, 105, pp. 311–12. From time to time, amateur astronomers become interested in the naked-eye visibility of Venus. Ellis' papers was followed by several letters to the Journal which are printed in subsequent issues.

Kiernan, N.S., 1985, 'Shadows Cast by Venus', *J.B.A.A.*, 95, p. 271. Venus casts shadow of observer in Delhi.

11.7 Where to look for a superior planet

Compare the entry of the superior planet into the morning sky from the eastern horizon to that of an inferior planet which emerges into the evening sky from behind the Sun at the western horizon.

11.8 Apparent changes in brightness of superior planets

Muirden, J., 1963, *Astronomy with Binoculars*, Faber. See pp. 61–2 on apparent changes in the brightness of Mars.

Juritz, C.F., 1910, 'On the Naked Eye Observations of Mars', *The Observatory*, 33, pp. 99–100. Juritz saw Mars from Cape Town, South Africa, with the naked eye two weeks after opposition. It was visible before sunset, 15° above the eastern horizon.

Chapter 12: Stars

12.1 Light without form

Delsemme, A., 1998, *Our Cosmic Origins: From the Big Bang to the Emergence of Life and Intelligence*, Cambridge University Press. A very thorough account of the formation of stars and planets and much else.

12.2 Where are the stars?

The distance between the Sun and the Earth is known to astronomers as the Astronomical Unit, or A.U. A parsec is about 200 000 A.U., and 5 parsecs is about one million A.U.

A light year is the distance that light travels in one year. The speed of light is 300 000 km/s, or approximately one billion kilometres per hour.

One light year is about 65 000 times the distance between Earth and Sun. 3.2 light years equal 1 parsec.

Use a planisphere to explore the changing orientation of the Milky Way to the horizon throughout the year.

There is a series of excellent illustrations of the relative positions of ecliptic and galactic plane in *The Astronomical Companion* by Guy Ottewell.

12.3 Star brightness

The term 'brightness' is often used to mean luminosity. Strictly speaking this is not correct: brightness is the subjective experience of luminosity. Nevertheless, it is possible to use the terms interchangeably where this does not lead to a clear misunderstanding. For more about luminance see Bohren, C.F., *What light*, chapter 15.

A star emits radiant energy across the entire electromagnetic spectrum. A cool star emits most energy at long wavelengths that are not perceived by the eye whereas a very hot star emits most energy at wavelengths that are shorter than can be perceived by the eye. A star may thus appear brighter in a photograph than it does to the eye because the eye has its peak sensitivity at shorter wavelengths than photographic emulsion. A star's brightness (or, more correctly, its luminosity) is therefore measured in a variety of ways. Astronomers allow for these differences by specifying the method used to arrive at a star's magnitude. Many stars are variable which means that their brightness changes perceptibly and periodically. Betelgeuse varies from about −0.1 to about 1.5 over some 5.7 years. A few stars vary in brightness over periods of several weeks.

Bobrovnikoff, N., 1984, *Astronomy Before the Telescope*, vol. 1, *The Earth–Moon System*, Pachart Publishing House, p. 67.

Padham, C.A. and Saunders, J.E., 1975, *The Perception of Light and Colour*, G. Bell and Sons, chapter 3, especially pp. 43–5.

Schaaf, F., 1988, *The Starry Room*, J. Wiley, chapter 10 on seeing faint stars.

12.4 Star colours

Compared with the eye, photographic emulsions are usually more sensitive to the blue end of the spectrum and less sensitive to the red end. For the eye it is the other way around.

Malin, D. and Muirdin, P., 1984, *Colours of Stars*, Cambridge University Press, chapter 1.

12.6 Why do stars twinkle?

Minnaert, M. *Light and Colour*. For shadow bands cast by Venus see section 12.6.

Young, A.T., 1971, 'Seeing and Scintillation', *S.&T.*, 42 (September), pp. 139–41.

12.7 Seeing in the dark

Bowen, K.P., 1984, 'Vision and the Amateur Astronomer', *S.&T.*, 67 (April), pp. 321–4.

Garstang, R.H., 1985, 'Visibility of Stars in Daylight', *J.B.A.A.*, 95, p. 133.

Henshaw, C., 1984, 'On the Visibility of Sirius in Daylight', *J.B.A.A.*, 94, pp. 221–22.

Hughes, D., 1983, 'On Seeing Stars (Especially up Chimneys)', *Q.J.R.A.S.*, 24, pp. 246–57.

Muirden, J., 1963, *Astronomy with Binoculars*, Faber, pp. 53–4.

Perelman, Y., 1958, *Astronomy for Entertainment*, MIR, p. 136.

12.8 Peripheral vision

Minnaert, M., *Light and Colour*, section 63.

Pirenne, M.H., *Vision and the Eye*, Chapman and Hall, pp. 57–8.

12.9 Why are stars star-shaped?

Minnaert, M., *Light and Colour*, section 59.

The Notebooks of Leonardo da Vinci, Oxford University Press, 1952, p. 54.

12.10 Constellations

Stars within a particular constellation are usually at vastly different distances from the Sun. Furthermore, they are usually moving in different directions to one another and so, over a period of thousands of years, the patterns they make will change.

Constellations of the (astrological) zodiac are, Aries (the Ram), Taurus (the Bull), Gemini (the Twins), Cancer (the Crab), Leo (the Lion), Virgo (the Virgin), Libra (the Scales), Scorpius (the Scorpion), Sagittarius (the Archer), Capricornus (the Sea Goat), Aquarius (the Water Carrier), Pisces (the Fish).

Chapter 13: Comets and meteors

13.1 Comets

Schechner, S.J., 1997, *Comets, Popular Culture and the Birth of Modern Cosmology*, Princeton University Press.
Edberg, S.J. and Levy, D.H., 1994, *Observing Comets, Asteroids, Meteors, and the Zodiacal Light*, Cambridge University Press, chapter 3 on comets.

13.2 Meteors

Meteors are also known as 'shooting stars' or 'falling stars', though, of course, meteors are not fragments of stars.

Bone, N., 1993, *Meteors*, Sky publishing Co. A guide to all aspects of meteors and their observation for the dedicated amateur.
Heide, F. and Wlotzka, F., 1995, *Meteorites*, Springer-Verlag. Deals with meteorites as opposed to meteors.

13.3 Artificial satellites

King-Hele, D., 1983, *Observing Earth Satellites*, Macmillan. The virtue of this book is that it is written specifically to get you outdoors and observing. Full of practical advice rather than theoretical discussions.

13.4 Aurorae

Bone, N., 1994, *The Aurora, Sun–Earth Interactions*, J. Wiley.
Davis, N., 1992, *The Aurora Watcher's Handbook*, University of Alaska Press. Everything you need to know about when and where to observe auroras.

'The Aurora Explained', set of two videos available from Armagh Planetarium, College Hill, Armagh, Northern Ireland, BT61 9DB.

13.5 Zodiacal light

Blackwell, D.E., 1980, 'The Zodiacal Light', in *Atmospheric Phenomena*, ed. D.K. Lynch, W.H. Freeman.

Edberg, S.J. and Levy, D.H., 1994, *Observing Comets*, chapter 6 on the zodiacal light.

McBeath, A., 1989, 'Zodiacal Light Complex', *Federation of Astronomical Societies Handbook*, ed. B. Jones, pp. 5–11.

Meinel, A. and Meinel M., *Sunsets*, chapter 10.

Olson, D.W., 1989, 'Who First Saw The Zodiacal Light?', *S.&T.*, 77 (Feb), pp. 146–8.

Appendix: Technical and practical advice for skygazing

Estimating distance.

Gatty, H., 1958, *Nature is Your Guide*, Collins. An outline of the many ways in which one can use natural phenomena to find one's way without using compasses or maps.

Binoculars

MacRobert, A., 1995, 'The Power of Binoculars', *S.&T.*, 89 (May), pp. 48–9.

A cloud primer

Schaefer, V.J. and Day, J.A., 1981, *A Field Guide to the Atmosphere*, Houghton Mifflin Co. Just about everything you could wish to know about clouds and rain and snow.

Glossary

Brody, B., 1994, 'Yellow', *P.T.*, 32 (April), pp. 220–1.

Lamb, T. and Bourriau, J., 1995, *Colour: Art and Science*, Cambridge University Press. Essays on various aspects of colour.

Nassau, K., 1983, *The Physics and Chemistry of Colour*, J. Wiley. Comprehensive and reasonably non-technical account of all the physical causes of colour.

Sekuler, R. and Blake, R., 1994, *Perception*. See chapter 6 on colour perception.

Padgham, C.A. and Saunders, J.E., 1975, *The Perception of Light and Colour*, G. Bell.

Index